The Pragmatist Challenge

The Pragmatist Challenge

Pragmatist Metaphysics for Philosophy of Science

Edited by

H. K. ANDERSEN AND
SANDRA D. MITCHELL

OXFORD
UNIVERSITY PRESS

OXFORD
UNIVERSITY PRESS

Great Clarendon Street, Oxford, OX2 6DP,
United Kingdom

Oxford University Press is a department of the University of Oxford.
It furthers the University's objective of excellence in research, scholarship,
and education by publishing worldwide. Oxford is a registered trade mark of
Oxford University Press in the UK and in certain other countries

First Edition published in 2023
Impression: 1

Published in the United States of America by Oxford University Press
198 Madison Avenue, New York, NY 10016, United States of America

British Library Cataloguing in Publication Data
Data available

Library of Congress Control Number: 2022945556

ISBN 978–0–19–880545–8

DOI: 10.1093/oso/9780198805458.001.0001

Printed and bound in the UK by
TJ Books Limited

For Joel, always. And to all my teachers and students,
who continue to inspire me.

—SDM

For A, E, and especially Endre, for encouragement, patience,
and great discussions.

—HKA

Contents

Acknowledgments

This volume emerged from ongoing discussions over several different workshops. While not all participants are represented in this volume, their questions and discussions helped shape the ideas that are here. In 2014, thanks to generous funding from Simon Fraser University's Department of Philosophy, the first workshop took place on the unceded traditional homeland of the Coast Salish First Nations with Sandra Mitchell, Huw Price, Jenann Ismael, John Dupre, Holly Andersen, and James Woodward, and faculty and MA students at SFU. The second workshop took place in 2015 at the University of Oslo, thanks to generous funding and facilitation of the Center for the Study of Mind in Nature. The speakers were Sandra Mitchell, Huw Price, John Dupre, Holly Andersen, James Woodward, Christopher Hitchcock, and Edward Hall. Thanks to Olav Gjelsvik and Bjorn Ramberg, and to Megan Delehanty and Paul Teller for involved work as discussants and moderators. The third instalment of the workshops took place in 2017, with thanks to the Center for Philosophy of Science at the University of Pittsburgh for their funding and facilitation. The speakers were Matthew Brown, Kerry Mackenzie, Laura Ruetsche, David Danks, Richard Healey, Jenann Ismael, Kenneth Waters, plus participants in the Center for Philosophy of Science community. The editors would like to thank Kesavan Thanagopal for proofreading assistance.

Andersen is grateful for the opportunity to live and work on the unceded traditional lands of the Squamish, T'seil Waututh, Kwikwetlam, and Musqueam First Nations. Mitchell wishes to recognize that the University of Pittsburgh occupies the ancestral land of the Adena Culture, Hopewell culture, and Monongahela peoples, and honor these traditional Native inhabitants.

List of Contributors

H. K. Andersen Philosophy, Simon Fraser University

David Danks Halıcıoğlu Data Science Institute & Department of Philosophy University of California, San Diego

Edward Hall Philosophy, Harvard University

Sandra D. Mitchell History and Philosophy of Science, University of Pittsburgh

Laura Ruetsche Philosophy, University of Michigan

James Woodward History and Philosophy of Science, University of Pittsburgh

1

Pragmatism for Philosophy of Science

H. K. Andersen and Sandra D. Mitchell

1.1 Introduction

This volume lays out a programmatic view for taking a pragmatist approach to topics in philosophy of science and metaphysics. Pragmatism involves a collection of specific views as well as comprising a general approach that can be applied to multiple topics. For topics at the intersection of philosophy of science and metaphysics, pragmatism as explored in this volume is an effective way to take entrenched debates and re-frame them in ways that move past old dichotomies and offer more fruitful paths forward. Each chapter explores a dual vision of pragmatism: specific pragmatist views are developed, demonstrating how to take a distinctively pragmatist approach to some particular issue or subfield; and the general shape of what it means to take a pragmatist approach is elucidated as well. The chapters thus tend to be synoptic in scope. Collectively, they offer a new approach that can be taken up in constructively reframing other discussions, ready to be applied to new specific topics. Pragmatism is an especially potent tool that sits at the interface between methodological and applied questions coming directly from sciences, on the one hand, and the underlying ontological or metaphysical commitments that are implied by or support the methodological discussions, on the other. The goal of the volume is to articulate a variety of ways to be a pragmatist without having to commit to a single specific set of -isms in order to make use of it, while highlighting the common themes that manifest across different discussions. The chapters offer a heterogenous yet programmatic approach to pragmatism.

There are classic topics in philosophy that have been the focus of discussion for centuries, even millennia, such as causation, modality, laws, and realism. The distinction between epistemology and metaphysics has often sent investigations of these topics down separate tracks, reflected in something

H. K. Andersen and Sandra D. Mitchell, *Pragmatism for Philosophy of Science* In: *The Pragmatist Challenge: Pragmatist Metaphysics for Philosophy of Science.* Edited by: H. K. Andersen and Sandra D. Mitchell, Oxford University Press.
DOI: 10.1093/oso/9780198805458.003.0001

of a schism between analytic metaphysics taking a more a priori approach, divorced from the details of messy, actual science, and philosophy of science which focuses more intensively on the actual practices, processes, and products of science. For example, the latter has moved from a syntactical approach to laws to something much more informed by the uses to which laws are put and the different ways in which claims about necessity, possibility, and actuality are formulated and realized in practice (Mitchell 1997, 2000). Pragmatism took many forms in the twentieth century (Føllesdal and Friedman 2006; Rescher 2013; Misak 2013), and continues to evolve, comprising a sometimes heterogeneous collection of ideas. Nevertheless, there are ongoing themes that recur with fresh faces over the more than a century in which pragmatism has been developing (Putnam 1995a and b; Longino 2004, 2018; Price 2011; Kitcher 2003; Chang 2004, 2022).

We suggest that both metaphysical and epistemological questions come to their sharpest form when considered in light of our most effective scientific practices. Questions about the best methodological processes for gaining knowledge are constrained by and emerge out of more fundamental questions about the nature of the world and our means of learning about it, and the principles required to connect the fundamental questions with the methodologies for answering them. There is a natural affinity between metaphysical considerations, epistemological methods, and the details of examples from contemporary science (e.g., Longino 1990; Dupré 1993; Magnus 2012).

Metaphysics and science have a long history together under the guise of natural philosophy. Natural philosophy is systematic in the sense of bringing together and justifying epistemological and metaphysical considerations as part of a package that inferentially hangs together. This way of approaching foundational questions uses metaphysical reasoning and principles as part of the well-oiled joints that connect other elements of the system. In like manner, the methodology derived from those broad considerations permits the acquisition of knowledge of nature to work effectively. Rejections of metaphysics as sterile and fruitless for human knowledge are historically almost as ubiquitous as defenses of metaphysics done in isolation. Rejections of metaphysics always presuppose a specific version of truth, reality, lawfulness, causation, and the like. The pragmatist challenge considered in this volume does not reject an a priori, analytic approach to metaphysics wholesale, but regards it as inadequate on its own to answer the questions it poses. Only by recognizing the functions and consequences of metaphysical claims on human action, and specifically epistemological practices, can we justify posits of what is true or what is real. We endorse a strategy of "doing metaphysics" from a human, non-ideal, perspective, that rejects a spectator theory of knowledge, and instead embraces the roles of human goals, scientific aims, and representational frameworks in deciding what there is in nature and how it works. This is the core idea of a pragmatist challenge: not a rejection of philosophical metaphysics, but offering an alternative approach to metaphysical inquiry that is closer in spirit to natural philosophy, closely integrating metaphysical and epistemological considerations.

The twentieth and twenty-first centuries witnessed the development of both analytic metaphysics and an analytic approach to the metaphysics of science. While the first is a priori, appealing to common sense intuitions, language, and logic, the Metaphysics of Science adopts the same philosophical methods but shifts the domain to scientific intuitions, language, and logic. Although Metaphysics of Science is concerned with key concepts that figure in science, its methods are those of the analytic philosopher (see, inter alia, Göhner and Schrenk 2019, Section 5). In contrast, the pragmatist challenge does more than just shift the territory for metaphysical analysis to science. It also expands that domain beyond the descriptive, representational features of scientific concepts and language to include wider practices of science. This includes not just how concepts like DNA or Higgs boson refer to what is real, but also how the processes and practices of experimentation, inference, and concept evolution give rise to and maintain the contingent "space of reasons" (Sellars 1963), which stabilizes reference and meaning. Importantly, by a focus on practice, the pragmatist challenge invokes different philosophical methods of analysis, reframing the usual metaphysical furniture of reality, truth, causes, laws, etc. in terms of their function in scientific practice, thus invoking norms for judgment that appeal to practical, experiential consequences. Metaphysics in this post practice-turn pragmatist variation has grown from roots in Kantian critical realism, empiricism, American Pragmatism, Sellarsian naturalism, and coherentism. Nature refracted by science simultaneously incorporates both the "objective" features with which science engages, but also the "subjective" features of the linguistic and non-linguistic scientific practices by which that engagement is enacted (Price 2011). It isn't just about what scientists say or must say, but also about what scientists do, or must do.

The workshops out of which this volume grew spanned a time period in which interest in pragmatist approaches has grown markedly. Since the first workshop in Vancouver in 2014, there has been increasing recognition of the centrality for analyses of science of concerns that might have been labeled purely metaphysical; and how much more effective metaphysical foundations can be when they fit into and are constrained by a recognition of the efficacy of actual, rather than hypothetical or idealized, science. We set up the first workshop to bring together a group of philosophers already known to be thinking in similar ways about these issues, to clarify how pragmatism works as an approach to metaphysical questions. How can progress be made on questions that arise in both contemporary metaphysics and philosophy of science by constructively reframing the way the questions are formulated? The initial plan was to use the workshop as a way of starting with a broad set of shared background assumptions to see how these could be used to launch further exploration of what it means to take a pragmatist approach to metaphysics and philosophy of science.

At the first workshop at Simon Fraser University (2014), organized by Holly Andersen and Sandra Mitchell, participants discovered that, while there was much philosophical commonality shared within that group of philosophers, we still identified many areas where we disagreed. No "Prolegomena to a Future Pragmatist Metaphysics" was in the offing. Instead, we continued to meet to explore the methods broadly construed as pragmatist. A second workshop (2015), organized by Holly Andersen, sponsored by the Center for the Study of Mind in Nature at the University of Oslo, extended the discussion, eliciting new views and considerations. At the third workshop (2017), at the Center for Philosophy of Science at the University of Pittsburgh, organized by Jim Woodward and Sandra Mitchell, participants focussed on both points of agreement and how a pragmatist strategy is applied in their own specific areas of study. We learned that if one is working in an area that emphasizes unity and realism, pragmatism is a useful tool to push back against specious unity; if one works in a field that tends to emphasize plurality and heterogeneity of methods, pragmatism provides a new way to articulate a vision of what underlying unity could nevertheless be coordinating that pluralism.

The chapters in this volume are thus not simply conference proceedings. Participants have had multiple rounds of presenting their own ideas, listening to others, getting challenged on particular points, and convincing others to shift their views. While not all those who participated in the workshops have papers in this volume, all the chapters have been presented in one of the workshops. These chapters thus represent the "what I'd like to say, after all that" version of authors' views. The result sets out a larger research agenda for applying a pragmatist approach to foundational issues.

1.2 What Makes the Challenge Pragmatist?

It's worth distinguishing between pragmatism as a philosophically coherent set of views, and pragmatism as it often is used in common parlance. In this volume, we've encouraged authors to use "pragmatist" to indicate a distinctively pragmatist idea. The term "pragmatic," or being pragmatic, is often used as a description for compromising on some otherwise desirable or normatively appropriate feature. Maybe we *want* better information before acting, but it would be cognitively expensive, so we decide to "be pragmatic," give up on the ideal to which we nevertheless aspire, and just do the best we can (such as bounded rationality in Simon 1957, 198 especially; philosophy for limited beings in Wimsatt 2007; and contemporary discussions around non-ideal epistemology, e.g., as highlighted and rejected by Begby 2021). This is often seen as a somewhat disappointing compromise: being pragmatic in this sense is to give up on a desire or goal that one legitimately wants but which would be hard to attain. It is to settle, to do with less. But this is not what it means to be a *pragmatist*: pragmatist and pragmatic are not interchangeable. While "pragmatic" as "settling" is consistent with and framed by a view of knowledge involving unattainable ideals, "pragmatist" involves rejecting those unattainable ideals in the first place, and rejecting the view of knowledge that involves striving for in-principle unattainable ideals to which one must resign oneself to never achieving. Pragmatism isn't giving up on some idealistic goal; it involves having a different goal.

It is also worth distinguishing pragmatics in the linguistic sense: being a pragmatist is not simply about language use. Pragmatism about truth or knowledge is not about pragmatics versus semantics as used in linguistics. Pragmatics adds information about the context of communication to semantics as part of studying word meaning, but pragmatism is not simply the study of word meaning, nor the addition of contextual information regarding communication. Pragmatism far outstrips pragmatics, and pragmatics can be non-pragmatist, in the sense of the philosophical approach known as pragmatism.

The pragmatist challenge is a working guideline for approaching disputes. It explores what is at stake in a debate between different camps offering apparently incompatible views or solutions on an issue such as Humeanism versus non-Humeanism about laws (e.g., Andersen 2018). The pragmatist challenge employs a strategy to push a debate about which claim is true, or which phenomenon is real, by getting serious about the attempt to answer, concretely, what difference does the answer make to experience? Pragmatism is distinguished from other approaches like verificationism or positivism by having a more encompassing view of what counts as relevance to experience. Attempting to answer the challenge of "what difference does it make?" helps to excavate shared presuppositions that might be implicit. It can help expose underlying dichotomies that are structuring the different views, and thereby provide a path forward that reframes or rejects the dichotomy underlying the different camps. The challenge is issued or taken up in order to be answered: it is not in-principle or meant to shut down discussion, but rather as an invitation to engage in new and different discussions. The chapters in this volume offer some examples of the pay-off for pragmatist thinking about epistemological and metaphysical issues in philosophy of science.

Taking up a pragmatist challenge is setting oneself the task to solve a philosophical problem through the use of the tool of pragmatism. The challenge part of a pragmatist challenge involves application of pragmatist ideas to the specifics of some area or topic within the sciences, doing the philosophical work to turn those broad themes into concrete proposals that fit effectively onto a subject. Many differences about what, exactly, a pragmatist should be committed to in order to count as a pragmatist turns on issues of context more than content. If a pragmatist can embrace unity in science, does that mean they thereby reject pluralism, and vice versa? What difference is made by adopting one view or another is contingent on the context. In the chapters that follow, the pragmatist approach pushes back against a dogmatic adherence to judging all decisions by what is warranted in one context. So, if a specific field has emphasized the need for One Right Theory, then a pragmatist can shift the emphasis to allow for a more heterogenous pluralism. Conversely, if a field has embraced pluralism to the extent that there is little left to unify the heterogeneity in the field, then a pragmatist can offer a view of the underlying unity at a more metaphysical level that sustains such pluralism in methods.

Even though pragmatism is sometimes associated with rampant pluralism, it can act as a path toward unification (e.g., Sellars 1956). Epistemological pursuits in the absence of genuinely metaphysical consideration may result in mere rote application of recipes, limiting our ability to understand how to appropriately and flexibly apply successful methods in new contexts. Metaphysical pursuits without epistemological constraints can drift into a vacuum, spinning wheels that move nothing, and leaving the discussion with insufficient resources to provide substantive ways to evaluate competing views. Requiring explication of the function or purpose of a distinction can clarify how to move forward (Hall 2020). There is thus an interesting way in which pragmatism can *reduce* a certain kind of pluralist proliferation by taking down the fences between subfields (Sellars 1956). Deploying the norm of the differences a view might make to experience, in the world, in empirical terms, allows us to sift through abstract metaphysical views identifying those which might otherwise differ only in which intuitions they codify and which can proliferate wildly but fruitlessly. Pragmatism can also offer more complex ways in which existing knowledge products can relate, beyond the traditional one of reduction (Danks 2018). A broader view of making a difference to experience can include taxonomization (e.g., Dupré 2001) and understanding (Achinstein 1984), in addition to well-known differences in measurable predictions. Asking what a method really requires the world to be like in order to be effective, and what implicit assumptions it involves that we can articulate explicitly, brings into view the hidden supporting infrastructure of methodology, revealing how the world must be for these methods to work, and providing us with the requisite understanding of those methods that allows for an informed, rather than rote, extension of the methods to new areas of inquiry (e.g., Hitchcock 2007; Woodward 2021). Pragmatism thus unifies methods by appeal to metaphysics, and culls metaphysical proliferation by appeal to epistemological constraints.

Pragmatism offers a way to accommodate more complex relationships than simple correspondence between linguistic and explanatory devices like models, diagrams, explanations, as products of science, and the parts of the world they were developed to represent, explain, and investigate. Pragmatism is well-known as founded in a rejection of a correspondence theory of truth; finding a way to replace that correspondence theory with something else is part of the pragmatist meta-methodology (e.g., Ramberg 2011; Teller 2012). Pragmatism brings together the collection of methods and attitudes that find value in embedding scientific practices in the context of human impact and goals, and not narrowly in representational fidelity of one particular sort (e.g., Massimi 2018). Truth can be an integral part of normative evaluation in the sciences when a different, pragmatist, view of truth is involved (e.g., Misak 2002, 2004). The metaphysical means is settled by the human ends in framing specific questions, set within the way the world is (e.g., Ismael 2016; Hitchcock 2012). The historical situatedness of current sciences is not to be eliminated as opposed to truth; it can be acknowledged while still improving our views and practices through time.

Debates within philosophy of science about topics such as truth, expertise, and weighing imperfect and potentially conflicting evidence have clear pragmatist import (e.g., Longino 2004; Staley 2007). The outward reaching ramifications of these issues within science have normative implications for how individuals living among others should guide their own behavior, making scientifically informed value-laden choices. How social institutions and practices can help policy should be mutually responsive with the sciences in an attitude toward mediating the exchange of scientific work with a broader public and policymakers, with social values and priorities, requiring clear communication in both directions (de Melo-Martin and Intemann 2016, 2018; Douglas 2009).

This volume investigates issues at the intersection where philosophy of science overlaps with contemporary metaphysics, by using pragmatism as a distinct way of answering metaphysical questions that differs from many contemporary analytic approaches to metaphysics. A pragmatist alternative to addressing metaphysical inquiries about foundational issues on which science bears and which are revealed by various sciences yields a different set of views, ones that tend to remain deeply immersed in the practices and products of the sciences, recognizing the historical situatedness of philosophizing without being tied to a particular moment in scientific history in a narrow way. Pragmatist challenges are not just to provide answers to what is true, what is fundamental, etc. Pragmatism also challenges the philosophical strategies for answering these questions, and the motivation for using those as the framing questions instead of questions that might lead in a different direction. Pragmatism as a tool in philosophical investigation can be applied to conceptual and methodological problems without specifying in advance what the eventual (pragmatist) solution will look like.

1.3 Pragmatist Themes in This Volume

A unifying feature of these chapters is the use of pragmatism as a methodological tool for approaching specific topics, rather than a specific kind of view on that topic. There are themes, ideas, or views that are characteristically pragmatist which recur in various chapters, though no single "here is what it means to be pragmatist" that all authors espouse. What pragmatism in a scientific context looks like has shifted a great deal since the early twentieth century, and many pragmatists disagree about specific pragmatist commitments. Fortunately, taking a pragmatist approach to specific topics does not require anyone to sign onto a specific list of philosophical commitments in order to count as pragmatist. It is a tool that can be put to many purposes. The way in which broadly pragmatist ideas are utilized will vary a great deal across different topics, based on the specific contexts in which they are deployed. Using a pragmatist approach to quantum theory (e.g., Ruetsche 2011; Jannson 2020) can yield a different emphasis in terms of realism than applying a pragmatist approach to systems biology (e.g., O'Malley and Dupré 2005), modeling practices for protein structure (e.g., Mitchell and Gronenborn 2017), and behavioral sciences (Longino 2012). The following are some of the more general themes from pragmatism in the twentieth and twenty-first centuries which appear in chapters in this volume.

Pragmatism is often associated with anti-realism, deflationism, or some kind of verificationism. Yet a pragmatist need not reject realism altogether (e.g., Dennett 1991; Putnam 1995b). The theme of balancing pluralism with unity and more modest forms of realism appear in the chapters by Ruetsche, Hall, Danks, and Mitchell. These chapters challenge the false dichotomy where either one must be anti-realist or there must be a single, unique, Right Answer that is unchanging through time to count as realist.

Hall and Danks offer different critiques of sharp or strong forms of realism, and come to surprisingly similar results—both trying to find this middle way in a sharply defined dichotomy with respect to realism (i.e., the correspondence or representation-only forms of realism).

For Danks, the issue is framed in terms of unification or unifiability. On his account, pragmatist philosophy of science can fall into four possibilities: realist philosophy of pragmatic science, pragmatic philosophy of realist science, realist philosophy of realist science, or pragmatic philosophy of pragmatic science. The question he poses in this framework: is pragmatism (in the philosophy of science) inconsistent with unified or unifiable science? He argues, contrary to some who endorse a pragmatist philosophy of science, that the answer is no. Danks detaches unification from strict derivational reduction, and offers a broader account in terms of hard and soft theoretical commitments.

Hall provides an analysis of what metaphysical or epistemological commitments are required by the philosophical use of terms like cause, or, the focus of his paper, law. He identifies three stances one might take. The first, robust realism, is where "law" has a role because it corresponds to something metaphysically real. On this view a law of nature is a fundamental metaphysical constraint on physical relationships, and "law" has a use because it refers to something metaphysically real. The second, modest eliminativism, is where "cause," for example, has no role in philosophical analysis, because it is not metaphysically real. Third is respectful deflationism, where such terms have a role in philosophical analysis for reasons other than being metaphysically real. Hall identifies respectful deflationism as pragmatist because those reasons that justify use of such terms are based on considerations other than correspondence to nature.

Danks and Hall both end up with a modest realism of sorts. It can be called deflationist in Hall's case, because the discussion to which his view is addressed is one that has highly emphasized a strong form of realism about such terms. Danks' urging of unification as compatible with pragmatism addresses a different kind of discussion that may have allowed for a proliferation of pluralism to which Danks draws attention to a role for underlying unification. And both are done in a pragmatist vein.

Ruetsche's chapter treats the content of arguably one of the most fundamental fields of physics, quantum field theories. She shows how QFT currently involves pragmatized content for interpreting the mathematical structure. She highlights the perennialist assumption that pragmatized content for a fundamental theory like QFT is temporary or in need of eventual elimination by something timelessly unchanging thereafter. She shows that the content is not pragmatized in a way that we should expect it to be reduced or eliminated; it will remain pragmatized. This means that whatever comes out of such a theory regarding what the world is fundamentally like, it is going to be ongoingly pragmatist in character. The pragmatism isn't a stand-in to get us by till the better, timeless, uniquely right interpreted content is found.

Ruetsche's critiques of the assumptions about timeless and unchanging knowledge also resonate with Andersen's view of truth as a process of trueing models and other knowledge products onto their targets. Keeping them in true, like the wheel of a bicycle, isn't something that is done once and finished forever. Knowledge products in science are dynamic through time, involving ongoing updating, refinement, and coordination.

This is highlighted in Mitchell's chapter as well, which examines the warrant for claims about realism. She identifies two fundamentalist strategies for metaphysical realism based on science, a top-down structuralist and a bottom-up causal strategy. She then breaks out of this dichotomy to develop an affordance-oriented interactionist metaphysics, where realism is warranted by stable phenomena that allow for the coordination in scientific practice of both the top-down and bottom-up strategies.

Several of the chapters offer a programmatic way of thinking with pragmatism to yield better ways of conceiving of veridicality and of usefulness of knowledge products that aren't tied to a narrow correspondence view of truth assigned atomistically to propositions. Mitchell, Danks, Woodward, and Andersen each touch on this.

Woodward offers a detailed version of functional analysis as a kind of means-end analysis. This puts at the center humans and their interests, highlighting how metaphysical questions can be made concrete by focusing on the purposes or goals we have in asking those questions. He distinguishes ambitious versus minimal metaphysics, aligning a pragmatist approach to philosophy of science as the latter. Functional analysis evaluates claims in terms of what role they serve in inferential contexts as a form of normative evaluation. He surveys how this functional analysis, as a specific way of being pragmatist, relates to many themes that recur in pragmatism. His synoptic survey includes usefulness and success as normative measures, the integral connections between description and evaluation of methodology, and how these can be applied to specific issues that are traditionally discussed in both metaphysics and philosophy of science.

Andersen develops a conceptual metaphor for truth that is suitable for philosophers whose work does not directly involve truth. She develops James' broadening view of experience for the pragmatist norm that to be meaningful is to make a difference to experience. Discussion around topics like idealizations would benefit from thinking about truth as like trueing a bicycle wheel, an ongoing process by which models are brought into true with their target system, judged by fit between belief and action. She concludes that arguing that we ought to be pragmatists about truth so that we can straightforwardly answer questions like "but is it really true that vaccines are effective?" without confusing caveats.

Collectively, the chapters here provide an alternative meta-metaphysics, a pragmatist approach to metaphysics. While this approach rejects many of the dichotomies in recent debates in metaphysics, it does not reject metaphysical inquiry. Rather, the pragmatist approach to metaphysics and science offers a positive new way to ask and answer these questions without the same presuppositions regarding the possibility of there being One Right Answer or some final end state to scientific inquiry that settles the questions for us. Pragmatism offers a way to settle disputes that tracks between one extreme of an overly simplified operationalism and the other extreme of a faith in some unknown and unknowable, potentially impossible, end of science as the basis for realism.

References

Achinstein, P. (1984, January). The pragmatic character of explanation. In *PSA: Proceedings of the Biennial Meeting of the Philosophy of Science Association* (Vol. 1984, No. 2, pp. 275–92). Philosophy of Science Association.

Andersen, H. (2018). A pragmatist challenge to constraint laws. *Metascience, 27*(1), 19–25.

Begby, E. (2021). *Prejudice: A study in non-ideal epistemology*. Oxford University Press.

Chang, H. (2004). *Inventing temperature: Measurement and scientific progress*. Oxford University Press.

Chang, H. (2022). *Realism for Realistic People*. Cambridge University Press.

Danks, D. (2018). Richer than reduction. In Danks, D., Ippoliti, E., eds., *Building theories. Studies in applied philosophy, epistemology and rational ethics*, vol. 41 (pp. 45–61). Springer.

de Melo-Martín, I., & Intemann, K. (2016). The risk of using inductive risk to challenge the value-free ideal. *Philosophy of Science, 83*(4), 500–20.

de Melo-Martín, I., & Intemann, K. (2018). *The fight against doubt: How to bridge the gap between scientists and the public*. Oxford University Press.

Dennett, D. C. (1991). Real patterns. *The Journal of Philosophy, 88*(1), 27–51.

Douglas, H. (2009). *Science, policy, and the value-free ideal*. University of Pittsburgh Press.

Dupré, J. (1993). *The disorder of things: Metaphysical foundations of the disunity of science*. Harvard University Press.

Dupré, J. (2001). In defence of classification. *Studies in History and Philosophy of Science Part C: Studies in History and Philosophy of Biological and Biomedical Sciences, 32*(2), 203–219.

Føllesdal, D., & Friedman, M. (2006). American philosophy in the twentieth century. *Daedalus, 135*(2), 116–26.

Göhner, J., & Schrenk, M. (2019). Metaphysics of science. *Internet Encyclopedia of Philosophy*. https://iep.utm.edu/met-scie/ accessed May 11, 2022.

Hall, N. (2020). Physical and metaphysical modality. In O. Bueno, S. A. & Shalkowski, eds., *The Routledge Handbook of Modality* (pp. 265–78). Routledge.

Hitchcock, C. (2007). Prevention, preemption, and the principle of sufficient reason. *The Philosophical Review, 116*(4), 495–532.

Hitchcock, C. (2012). Events and times: A case study in means-ends metaphysics. *Philosophical Studies, 160*(1), 79–96.

Ismael, J. (2016). How do causes depend on us? The many faces of perspectivalism. *Synthese, 193*(1), 245–67.

Jansson, L. (2020). Can pragmatism about quantum theory handle objectivity about explanations? In S. French & J. Saatsi, eds., *Scientific realism and the quantum* (pp. 147–67). Oxford University Press.

Kitcher, P. (2003). *Science, truth, and democracy*. Oxford University Press.

Longino, H. E. (1990). *Science as social knowledge: Values and objectivity in scientific inquiry*. Princeton: Princeton University Press.

Longino, H. E. (2004). How values can be good for science. In P. Machamer & G. Wolters, eds., *Science, values, and objectivity* (pp. 127–42). University of Pittsburgh Press.

Longino, H. (2012). Knowledge for what? Monist, pluralist, pragmatist approaches to the sciences of behavior. In K. Plaisance & T. Reydon, eds., *Philosophy of Behavioral Biology* (pp. 25–40). Springer.

Longino, H. E. (2018). *The fate of knowledge*. Princeton University Press.

Magnus, P. (2012). *Scientific enquiry and natural kinds: From planets to mallards*. Springer.

Massimi, M. (2018). Four kinds of perspectival truth. *Philosophy and Phenomenological Research*, *96*(2), 342–59.

Misak, C. (2002). *Truth, politics, morality: Pragmatism and deliberation*. Routledge.

Misak, C. J. (2004). *Truth and the end of inquiry: A Peircean account of truth*. Clarendon Press.

Misak, C. (2013). *The American pragmatists*. Oxford University Press.

Mitchell, S. D. (1997). Pragmatic laws. *Philosophy of Science, 64*, S468–S479.

Mitchell, S. D. (2000). Dimensions of scientific law. *Philosophy of Science*, *67*(2), 242–65.

Mitchell, S. D., & Gronenborn, A. M. (2017). After fifty years, why are protein X-ray crystallographers still in business?. *The British Journal for the Philosophy of Science*, *68*(3), 703–723.

O'Malley, M. A., & Dupré, J. (2005). Fundamental issues in systems biology. *BioEssays*, *27*(12), 1270–6.

Price, H. (2011). *Naturalism without mirrors*. Oxford University Press.

Putnam, H. (1995, January). Pragmatism. In *Proceedings of the Aristotelian Society* (Vol. 95, pp. 291–306). Aristotelian Society, Wiley.

Putnam, H. (1995). The many faces of realism. In R.B. Goodman, ed., *Pragmatism* (pp. 163–81). Routledge.

Ramberg, B. (2011). Method and metaphysics: Pragmatist doubts. In J. Malpas, ed., *Dialogues with Davidson: Acting, Interpreting, Understanding* (pp. 129–46). The MIT Press.

Rescher, N. (2013). *The pragmatic vision: Themes in philosophical pragmatism.* Rowman & Littlefield.

Ruetsche, L. (2011). *Interpreting quantum theories.* Oxford University Press.

Sellars, W. (1963). *Science, perception, and reality.* New York: Humanities Press.

Sellars, W. (1956). Empiricism and the philosophy of mind. *Minnesota Studies in the Philosophy of Science, 1*(19), 253–329.

Simon, H. A. (1957). *Models of man; social and rational.* Wiley.

Staley, K. W. (2007). Evidential collaborations: Epistemic and pragmatic considerations in "group belief." *Social Epistemology, 21*(3), 321–35.

Taylor, P. C. (2004). What's the use of calling Du Bois a pragmatist?. *Metaphilosophy, 35*(1–2), 99–114.

Teller, P. (2012). Modeling, truth, and philosophy. *Metaphilosophy, 43*(3), 257–74.

Wimsatt, C. (2007). *Re-engineering philosophy for limited beings: Piecewise approximations to reality.* Harvard University Press.

Woodward, J. (2021). *Causation with a human face: Normative theory and descriptive psychology.* Oxford University Press.

2
Sketch of Some Themes for a Pragmatist Philosophy of Science

James Woodward

2.1 Introduction

This chapter sketches, in a very partial and preliminary way, an approach to philosophy of science that I believe has some important affinities with philosophical positions that are often regarded as versions of "pragmatism." However, pragmatism in both its classical and more modern forms has taken on many different commitments. I will be endorsing some of these and rejecting others—in fact, I will suggest that some elements prominent in some recent formulations of pragmatism are quite contrary in spirit to a genuine pragmatism. Among the elements I will retain from many if not all varieties of pragmatism are an emphasis on what is useful, where this is understood in a means/ends framework, a rejection of spectator theories of knowledge and skepticism about certain ways of thinking about "representation" in science; also skepticism about ambitious forms of metaphysics. Elements found in some previous versions of pragmatism that I will reject include proposals to understand (or replace) truth with some notion of community assent and skepticism about causal and physical modality. It is possible that the particular collection of ideas I will advocate is sufficiently idiosyncratic that it is held only by me and no one else. But in the end, I don't much care whether the views I will be describing count as an instance of "pragmatism" or something else; what I do claim is that they have some internal coherence and some motivation in ideas found in writers like Charles Sanders Peirce, William James, John Dewey, and more recently, Hitchcock, Mitchell, Price, and Wilson among others.

James Woodward, *Sketch of Some Themes for a Pragmatist Philosophy of Science* In: *The Pragmatist Challenge: Pragmatist Metaphysics for Philosophy of Science.* Edited by: H. K. Andersen and Sandra D. Mitchell, Oxford University Press.
DOI: 10.1093/oso/9780198805458.003.0002

2.2 Orienting Remarks

First, some orienting remarks. As I see it, philosophy of science—and in particular what is sometimes called general philosophy of science—is under pressure as a discipline. On the one hand, the interests of many philosophers of science have shifted strongly to philosophy of the specific sciences and there is general acceptance of the idea that to do philosophy of science credibly one has to know a substantial amount about one or more particular scientific discipline. Learning such science can be a great deal of fun but there are dangers. One is that the more specialized philosophy of science becomes, the more fragmented it becomes. In consequence, it becomes more and more difficult for these different specialists to talk to one another, not to mention interact meaningfully with the rest of philosophy. And like it or not, philosophy is going to continue to be the discipline in which most philosophers of science will get jobs for the foreseeable future.[1]

Another danger with an exclusive focus on philosophy of the particular sciences is that in practice this sometimes leads to products that look like "mere science reporting" or "science journalism"—virtually all of the philosopher's energy goes into mastering some cool body of science and describing it to philosophical audiences. The distinctive contribution of the philosopher—the intellectual "value added"—looks rather minimal, even if we agree that there is some value to knowledge transfer and intellectual arbitrage. When I speak of "the distinctive contribution of the philosopher" I do not mean that this needs to look like philosophy, traditionally understood. I mean only that it would be desirable for the philosopher to add something that is intellectually valuable in addition to the science that is reported. This might be something distinctively philosophical but if not that, it might, for example, consist of an analytical framework that allows one to see the science or methodological issues surrounding it in a new light.

However, if we look to so-called general philosophy of science (as currently practiced) as one possible source of tools or ideas that might be

[1] Just to clarify: I do *not* mean by this remark to suggest that there is necessarily something wrong or unfortunate about philosophers of science being primarily located in philosophy departments. In fact, I think that philosophy of science as a discipline would benefit from more contact with the rest of philosophy and particularly from the influence of standards of argument and precision found in the best "non-science" philosophy. I just mean to draw attention to the tension between, on the one hand, learning a lot of science and scientific methodology and, on the other hand, keeping up with developments in philosophy. I see the current interest in metaphysics of science as one response to this tension: the philosopher of science reconnects with philosophy by doing metaphysics.

employed in providing this value added, the results seem, with some conspicuous exceptions, underwhelming. The grand narratives associated with Popper, Kuhn, and Lakatos seem stale and played out and in any case were largely pitched at far too high a level of abstractness and generality (and assumed far too much uniformity in the practice of science) to be useful in the present context. The analytical tradition associated with writers like Hempel is also in many respects unsatisfying, in part because of its lack of engagement, at least in many respects, with the content of real science—there is too much first-order logic, theory T, and "All ravens are black" and similar stuff. (I acknowledge that I'm caricaturing—for example Hempel himself had interesting things to say about the philosophy of the social sciences and psychiatry. However, caricatures can convey truths.)

Enter the siren song of metaphysics: Perhaps philosophers of science should do the metaphysics of science (or they should replace what they are doing with the metaphysics of science or what they were doing all along was metaphysics of science, often badly, but they can do it better by adopting the ideas of specialists in metaphysics). An apparent attraction of this suggestion is that it directly addresses issues about how what philosophers of science are doing differs from science reporting (philosophers of science are doing some variety of metaphysics) and it provides an avenue for restoring connections with portions of the rest of philosophy. But what should philosophers of science do if they want to preserve some of the concerns that animate general philosophy of science, but also don't want to do metaphysics of science, at least as this as currently practiced by many analytic metaphysicians?[2]

[2] Some may wonder: why try to do "general" philosophy of science at all, rather than philosophy of the more specific sciences? There are a number of possible responses in addition to those already gestured at (disciplinary fragmentation, connection with the rest of philosophy etc.): (1) issues often thought of as within the province of general philosophy of science—having to do with e.g. evidence and theory testing, causal reasoning, explanation, inter-theory relations theory structure, and so on are increasingly discussed in interesting and sophisticated ways in other disciplines such as statistics, artificial intelligence, and psychology. General philosophy of science can (and has) contributed to this discussion and doing so enables it to retain contact with these other disciplines. That the other disciplines are concerned with these problems suggests both that they are real and that it is possible to make progress on them, even if philosophers have not always succeeded in doing so. (2) (Here I am going to step on some toes) philosophers of science should ask themselves where their comparative advantage lies—what special skills and background they have that best enables them to contribute to the general intellectual and scientific culture. I don't deny that some philosophers of discipline X may be in as good a position to contribute to X as researchers whose professional home is within X itself, but this does not seem to me to be common. In many cases the abilities and background of philosophers of science are better suited to general philosophy of science.

2.3 Minimal vs. Ambitious Metaphysics

At this point some clarifications/distinctions are needed. "Metaphysics" as currently practiced is a very heterogeneous activity. At an extremely abstract level, one might think of metaphysics (or ontology) as having to do with what is "out there" or what "exists." Understood in this way, the claim that there are electrons is (I suppose) a metaphysical claim. If one holds, as I do, that causal relations hold in nature, independently of our projective activities, this is again a metaphysical claim. By the same standard, the denial of this claim, according to which causal relations are not out there but rather something that we project onto nature is also metaphysics. Of course, this understanding of the subject effectively makes it impossible to avoid doing metaphysics. In any case, claims about the reliability of various inference procedures, which I take to be one central concern of a pragmatist philosophy of science (see below), rest on claims about what is out there in nature; such procedures are reliable to the extent that they discover *truths* (or near truths) about the world. Similar commitments are present when one attempts to *explain* why various procedures and patterns of thinking are reliable since this inevitably involves appeal to the presence of worldly structures that support such procedures and patterns. However, according to the version of pragmatism that I favor, these facts about what the world is like are ordinary, contingent empirical facts that are discoverable by science. One might argue about whether claims about such facts are usefully described as "metaphysics," but assuming one decides to so characterize them, I suggest that we use the phrase "minimal metaphysics."

I contrast minimal metaphysics with metaphysics of a more ambitious or expansive variety. I won't try to provide necessary and sufficient conditions for the latter but here are some diagnostic features: (1) Appeals to special entities and relationships that seem to play no role in science or in ordinary empirical discourse such as "relations of necessitation between universals" claimed to underlie and explain facts about the obtaining of empirical regularities (Armstrong 1983). (2) Claims that in addition to ordinary scientific explanations, "metaphysical explanations" are also required to "ground" various scientific or ordinary empirical claims or practices, as with Loewer's (2012) claim that we need a metaphysical explanation of laws of nature, where this is explicitly contrasted with ordinary scientific explanation. (3) "Reconstruction" projects that have no obvious motivation internal to science but are instead motivated by extra-scientific concerns. A prominent example is Lewis' (1986) attempts to reduce laws of nature to facts about patterns in a Humean mosaic consisting of facts about the instantiation of intrinsic, perfectly natural properties at spacetime points. (I claim there is no obvious scientific reason for attempting to do this.) (4) Worries about what "really" exists, where the modifier "really" imports some non-ordinary standard for existence, as when it is claimed that tables do not really exist, and all that really exists are atoms arranged table-wise.[3]

My point in introducing this contrast is not to argue (at least here) against expansive metaphysics but rather to urge that it is different from minimal metaphysics and that a pragmatic philosophy of science needs only the latter and not the former. My aim in what follows (and I apologize that this sounds so grandiose and pretentious) is to sketch, in a very preliminary way, some fragments of a picture of what philosophy of science might be or what philosophers of science might do, that relies only on minimal and not ambitious metaphysics but also that does not return us to the bad old days of theory T and "All ravens are black."

As already suggested, my organizing idea is that this will take the form of a *pragmatist* philosophy of science. This formulation is intended to highlight the twin projects of (i) *description/interpretation* and (ii) *methodology* as a normative enterprise that is central to philosophy of science. Insofar as these were always concerns of general philosophy of science, it attempts to retain some continuity with and to revivify this enterprise. As suggested below, one way in which such a connection with general philosophy of science might be retained is through a focus on common methodological problems and issues that occur across different scientific disciplines. For example, there are common methodological problems concerning causal inference from observational data that recur across disciplines as different as economics and neurobiology. These are natural foci for generalist philosophers of science. Investigating these problems requires a lot of attention to specific details—thinking in terms of deducing observation O from

[3] An inevitable response to this distinction will be that it is less clear and sharp than I suggest. I won't try to fight this battle here. I do claim (i) that the distinction is analytically defensible and (ii) it corresponds to something sociologically "real." Papers in the *Physical Review* contain (at most) minimal metaphysics and are usually not written by philosophers. Papers in *Oxford Studies in Metaphysics* are written by philosophers, often contain ambitious metaphysics, and are likely not publishable in the *Physical Review*. There is a difference here that we can recognize, even if there are borderline cases.

I also recognize that one way of attacking the minimal/ambitious distinction is to contend that when we try to "interpret" scientific theories we find that many make ambitious metaphysical claims and that these are supported by the empirical success of those theories—we can read off ambitious metaphysics from the science. My reasons for skepticism about such contentions can be found in Sections 2.5–2.9.

hypothesis H is an unhelpful level of abstraction. Nonetheless at least some of the relevant details are common across different disciplines. So in this respect we have "generality" (as in "general philosophy of science") but not the kind of generality associated with theory T and HD models of confirmation. And we have a kind of engagement with details that (because it involves focusing on problems that recur across different areas of science) is different from the engagement that is present if one focuses only on the philosophy of some specific science such as biology.[4]

A second, related way in which some measure of generality might be retained is through a focus on developing analytical tools that allow for perspicuous description/interpretation of specific strategies, structures and argument patterns common to different areas of science. For example, a common problem in many areas of science from physics to biology arises from the use of a plurality of models in these disciplines that capture features of systems of interest at different scales (of length, time, energy) but where the relations or connections of these models are often very unclear. Classic Nagelian models of reduction seem descriptively inadequate and unhelpful for understanding such relationships. Scientists working on problems of relating models at different scales use language like "getting models at different scales to talk to one another" or "to exchange information" but these are just metaphors—a worthwhile task for philosophers of science would be to develop crisper tools for describing model relationships in such cases. Once we have more appropriate descriptive tools and a better understanding of the variety of strategies employed in different areas of science to connect models at different scales, we may also be in a better position to offer methodological advice, since, depending on empirical details, some connecting strategies are more likely to be successful than others.[5] Again,

[4] In other words, if one wants to engage with science in a detailed way, there are at least two different ways of doing this. One is to divide up the territory in terms of currently recognized scientific disciplines: philosophy of physics, biology etc. The other way focuses on specific problems that may be common to a number of different scientific disciplines. I recommend that some philosophical work engage with the latter.

[5] For example, philosophical proposals regarding the best way to understand the relationships between psychology and neurobiology have typically considered only a very small range of possibilities—Nagelian type-type reductions, supervenience relations with token identities and so on. But there are a huge number of other possibilities regarding relations between theories and model at different "levels," instantiated in other areas of science, that are arguably better models for the psychology/neurobiology relationship. I intend "better" here has a normative as well as a descriptive component—e.g. we are more likely to discover informative relationships between psychology and neurobiology if we do not look for type-type identities.

this involves much more than mere science reporting, but the results need not involve ambitious metaphysics.[6]

The remainder of this chapter is organized as follows. Section 2.4 sets out some overall themes and commitments of a pragmatist philosophy of science. Section 2.5 discusses the contrast, developed by Huw Price, between subject and object naturalism and its relevance to philosophy of science. Sections 2.6–2.7 discuss and illustrate the role of means/ends reasoning in a pragmatist philosophy of science. Section 2.8 discusses representation and Section 2.9 realism versus instrumentalism from a pragmatist perspective. Sections 2.10 and 2.11 take up some additional issues having to do with the "ontological commitments" of scientific theories and the status of modal claims.

2.4 Some Themes

2.4.1 Usefulness, Utility, and Success

I take pragmatism to involve a focus, first of all, on usefulness and utility, on what "works" or does not work. Applied to science, this involves a commitment to the idea that at least sometimes, scientific investigation works *successfully* to deliver information of various sorts about nature. The pragmatic philosophy of science I favor seeks to understand how this is possible—what sort of information is so delivered, what the methods and strategies are that are involved in its production, and how it is that these methods are successful to the extent they are, which in turn requires understanding the structures in nature (again understood in terms of minimal metaphysics) that support such success. To adopt a common characterization, this involves an *instrumental* approach to science—instrumental in the sense that theories/models and various scientific practices of reasoning and intervening are seen as instruments or tools for the achievement of various purposes and evaluated accordingly. However, as we shall see, this does *not* involve a commitment to *instrumentalism* in the sense in which philosophers understand this notion.

[6] Examples of work of the sort of work I have in mind include Batterman (2013) and Wilson (2017).

2.4.2 Description and Evaluation

As I see it, pragmatism as applied to philosophy of science has both a descriptive or interpretive component—characterizing accurately relevant aspects of scientific practice and reasoning—and an evaluative or normative component: we want to understand why various methods and strategies are successful in delivering knowledge (why they "work"), to the extent that they do, but also to identify cases in which present methods and strategies fail to deliver what they claim to deliver and how, in the light of this, these might be improved. Thus critical evaluation of methods and approaches currently employed in various areas of science (and not just description of these) is an important element in the normative part of this enterprise. I see these as falling under the general rubric of *methodology*,[7] which is central to pragmatist philosophy of science as I conceive it.

2.4.3 Means and Ends

The enterprise described in Sections 2.4.1 and 2.4.2—and particularly its normative component—is naturally pursued within a means/ends framework: Investigators have certain ends or goals that are part of the scientific enterprise. A very partial list includes but is not limited to successful prediction, causal analysis, and explanation, exploring the mathematical structure of various theories and models, description and classification for specific purposes, and building and making things.[8] (These goals may be further subdivided—e.g., there are a number of different possible predictive problems with different structures.) Investigators employ various means (methods, strategies, etc.) for achieving those goals. These means can either be well-adapted to or conducive to these goals or not and both description and evaluation should be pursued in the light of this. In other words, normative assessment proceeds by asking whether various possible means are effective in reaching our goals, with ineffective means being discarded in

[7] It is sometimes claimed that there is no such thing as "the scientific method." If the quoted phrase refers to a single method for all of science at the level of generality sought by philosophers like Popper and Lakatos, I agree. But this is compatible with their being many more specific methods employed in particular sciences. I see these as belonging to methodology—see below—and very much within the purview of philosophers of science.

[8] I emphasize that, although, as will become apparent below, I regard the discovery of truths about nature as among the goals of the scientific enterprise, science has goals (e.g. building) that don't reduce to the discovery of truths.

favor of more effective ones. Means/ends analysis is discussed in more detail in Sections 2.6–2.7 but examples of this sort of approach can be found in much of classical (frequentist) statistics, in formal learning theory, as illustrated by work of Kelly (e.g., 2007) and Schulte (2017), in machine learning (for a causal inference application see Janzing et al. 2012) and (in connection with issues having to do with variable choice) in Hitchcock (2012) and Woodward (2016).

Features of scientific inquiry that can be profitably understood within a general means/ends framework include: procedures for hypothesis-testing and inferring from evidence to different sorts of hypotheses, issues having to do with choice of variables or vocabulary for framing theories, many issues concerning the structure of scientific explanation and causal analysis, and many aspects of modeling practices, including strategies for "backgrounding" or "neglecting" certain features of systems in order to better understand others, as well as the idea that models can often function so as to provide information about nature, while not always functioning so as to mirror or picture nature.

Thus, a central part of the pragmatist program is to apply the means/ends framework at many different levels and to many different problems in philosophy of science. The picture is thus pragmatist twice over—science itself is thought of as a pragmatist, means/ends enterprise and we use this pragmatist framework to understand and evaluate it. What results is not a single grand narrative or single overarching treatment of science in the manner of Popper, Kuhn, or Lakatos. Rather, various particular and specific practices of inference, modeling and so on are understood with a means/ends framework. For this reason, it is hard to fully convey what this approach involves in a brief summary; rather whatever value or insight the approach provides requires illustration on a more piece-meal, case by case basis—in connection with the evaluation of particular strategies and practices pursued in specific empirical contexts.

2.4.4 Description and Evaluation Together: An Analogy

As I have been emphasizing, on this approach to science description and evaluation should work together. As a partial illustration/analogy for how this might work, consider theories of human vision as developed by vision scientists. (This analogy is due to Alison Gopnik.) Human beings form beliefs about aspects of their environment on the basis of visual input and vision scientists wish to understand the processes and mechanisms that underlie this. But human beings do not just form beliefs on the basis of visual inputs, they are often pretty good at forming reliable beliefs on this basis—that is to say, the visual system is fairly successful (within certain limits) in providing accurate information about the visual environment (which is not to say that the visual system is in the business of constructing an internal representation which is isomorphic to every feature of the environment, a point that should be kept in mind when thinking about how scientific models incorporate information about nature). Part of what vision scientists want to explain is this success—what it consists in, how it is possible. Doing so has often taken the form of formulating "ideal observer" or other sorts of normative theories of visual processing—theories that look to understand how (that is, in virtue of what input information and subsequent processing) it is possible to extract information that in at least some respects is fairly reliable about the visual environment. One tries to formulate both high level and much more specific goals that the visual system or components of it have and then tries to understand the means by which the system achieves these goals (to the extent that it does)—for example, the computations that the visual system employs and the algorithms that realize these computations. The result is both a descriptive account of the operation of the visual system and a kind of evaluation in the sense that it provides material for assessing how well the system does in achieving these goals. The normative theory or theories furnish a framework or benchmark for understanding what is going on descriptively. They also can provide an explanation for why the visual system succeeds to the extent that it does. Note that the source of normativity here (as with the other examples I have mentioned) has to do with assumptions about ends or goals; the end of providing accurate enough information about the visual environment provides the basis for assessing the computations that the visual system employs in achieving this end.

In recent years some psychologists have adopted a similar sort of approach to understanding more cognitive activities including causal cognition. Given certain goals associated with causal cognition, one can formulate "rational learner" accounts which explore which strategies of learning from evidence and of reasoning will conduce or best conduce to those goals. One can then investigate empirically whether and to what extent various kinds of subjects—human adults, small children, non-human animals—follow those strategies and procedures and to what extent they do not. The result is both a descriptive account of aspects of causal cognition and an account that has an evaluative or normative component. This sort of work has been carried out by psychologists like Gopnik, (e.g., Schulz et al. 2007), Patricia Cheng (Carroll and Cheng 2010), and Josh Tenenbaum, among others.[9]

What I am recommending is that a pragmatic approach to philosophy of science should follow a similar path.

2.4.5 More on Evaluation

My emphasis on methodology and the normative element in a pragmatic philosophy of science may be disquieting in the sense that it recalls another feature of the bad old days—the willingness of some philosophers of science to dismiss large swaths of scientific theory on the grounds that these violate favored a priori methodological principles. (Recall, for example, Popper's dismissal, at least at one point, of evolutionary theory as unscientific because unfalsifiable, Hempel's insistence that explanations in biology, the social sciences and history should ideally appeal to "laws of nature"—a requirement rarely if ever met in these disciplines—and claims that theories that appeal to "unobservables" are unscientific or at least should be replaceable in principle by theories that do not have this feature.) What I have in mind is different in several respects. First, the sort of methodological inquiry I have gestured at above involves hypothetical rather than categorical imperatives. They take the form: if you want to achieve goal G in circumstances C, you should employ means X and not means Y, where this leaves open the possibility that some different means should be employed in circumstances different from C. (If the empirical circumstances characterizing biological investigations are different from those characterizing investigations in high energy physics, as they clearly are, different methods may be appropriate.) By contrast a priori methodological requirements are usually understood as a categorical and (if any attempt to is made to justify them) they are often taken to follow from the nature of "rationality" or to be requirements of "logic" in some broad sense. Alternatively they are taken to be vindicated by some form of rational intuition.[10] As a result, the

[9] Work of this sort on causal cognition is discussed in Woodward, 2021a.

[10] Consider the ubiquitous references to "logic" in various influential programs in mid-twentieth century philosophy of science: the logic of confirmation, the logic of explanation, and so on. Exactly what "logic" means in these contexts is not so clear, but one consequence of this form of framing is that means ends/analysis is not employed—criteria for hypothesis

requirements are understood as holding independently of empirical circumstances and as not subject to ordinary means/ends assessment.

The hypothetical imperatives relevant to methodology can be established in a variety of ways including mathematical demonstration, heuristic argument, simulation, and empirical investigation (see below, Sections 2.6–2.7). For example, in some cases one can prove, mathematically, that if one wants to achieve certain goals, certain means will lead to those goals and others will not—this is illustrated by the choice of an estimating procedure with good error characteristics discussed in Section 2.7.

As an empirical matter, methodological criticism and advice seems more appropriate (and needed) in some areas of science than others. Pragmatist philosophers should thus focus on the most appropriate areas. Perhaps (I'm not in a good position to say) physicists have their house in order, methodologically speaking. But in areas of science in which I am most interested—e.g., problems of causal inference and reasoning in the social, behavioral and bio-medical sciences—there are uncontroversial arguments (some of them of a mathematical nature, some of them quasi-empirical) that sub-optimal or fallacious methods are employed with some frequency.[11] Pragmatist philosophers of science should not be afraid to draw attention to this—indeed this (along with suggestions for improvement) is one of the primary ways they can be useful.

2.4.6 The Centrality of Action and Interaction with the World in Science and in Knowledge Acquisition More Generally

A common although far from universal theme among many pragmatists, emphasized particularly by Dewey, is a distaste for "spectator" theories of knowledge. We don't just passively observe the world, we often can *act* on it, changing and manipulating it. It is not just that the world causally affects us—e.g., through perception; we causally affect the world. This matters a

"confirmation" are not advocated on the means/ends basis that following them will lead to some desirable outcome like truth but rather have some other basis—perhaps "intuition" or ease of analysis with the tools of first-order logic.

[11] A simple illustration is provided by the common practice of using regression equations as a general method for inferring causal relations from statistical data and "controlling" for everything the researcher can think of in estimating these equations. This demonstrably leads to incorrect results for many reasons including "collider bias." See Pearl (2018) for a discussion of this example and many other illustrations of bad but common statistical practice. Contemporary economics is also fertile ground for informed philosophical criticism.

great deal for the practice of science. In the context of philosophy of science, pragmatists will thus attach a great deal of importance to the role of experimentation and to activities of making or constructing or building (making includes everything from the construction of instruments to synthesis of new chemical compounds or in the case of social science, the creation of new institutions). These activities are legitimately part of science and are not to be dismissed as of "merely pragmatic" significance or as unimportant because they are only part of "applied" (as opposed to "pure") science. Experimentation, construction, and applications of science should be thought of sources of information about the world that are incorporated into and help to structure our models and theories—the results of such manipulative activities tell us something about what the world is like. Thus, as good pragmatists, we should resist putting "merely" in front of "pragmatic" and also resist sharp distinctions between "pure" and "applied" science—more on this below.[12]

2.4.7 Modal Knowledge

The role of action and associated notions having to do with manipulation, control, planning, and so on is particularly important in gaining an adequate understanding of knowledge involving physical modality in science—knowledge having to do with causal relationships, with physical possibility and necessity, and with counterfactuals. There is a tendency among some philosophers sympathetic to pragmatism (e.g., Sellars, perhaps Brandom) to attempt to understand notions connected to physical modality in terms of a spectator picture, so that, for example, laws and causal generalizations are understood as "inference tickets" or as "licenses" for predicting the value of one variable for another. Laws and causal generalizations do play this role but they are also centrally bound up with concerns having to do with manipulation and control and their role in science cannot be understood apart from these. Relations that support inference and prediction but not manipulation (such as the relationship between barometer readings and storms) should be distinguished from relationships that support manipulation and control. Physical modality has to do with the latter.

[12] More generally of course it goes without saying that pragmatist philosophers of science will not think that labeling a consideration as "pragmatic" is good grounds for dismissing it.

I believe that spectator approaches to modality tend to encourage either (i) a tendency toward misplaced reification (with modal claims taken to refer to special sorts of objects or properties—see Section 2.11 below) or, alternatively (ii) skepticism or various forms of subjectivism about modal notions. A paradigm of (i) is Lewis's treatment of modality in terms of "possible worlds," each as real as our own. Illustrations of (ii) include the view that modal claims are unclear or not part of legitimate science (a position adopted by, for example, Quine) or that they are rooted in facts about our projective or epistemic organizing activities, rather than how matters stand in the world. I believe that both alternatives (i) and (ii) become much harder to maintain when one thinks of modal knowledge as rooted in activities like experimentation. How, for example, are we to understand the characteristic concerns of experiment design if the sort of modality involved in causal claims is understood along the lines of (i) or (ii)? When we do a randomized experiment designed to assess the efficacy of a drug, and reach a conclusion about the causal role of the drug in promoting recovery, it does not seem plausible that we should understand this either as an inquiry into a possible world distinct from the actual world or as an inquiry into our own projective proclivities.

2.4.8 Spectator Orientation as Rejected by Some but Not All Self-Identified Pragmatists

I said that a distaste for spectator theories of knowledge is common but far from universal among pragmatists. I agree with Peter Godfrey-Smith (2015) that at least two recent philosophers who are often regarded as pragmatists—Quine and Rorty—seem to be in the grip of this spectator conception. Quine, for example, talks of sense experience "impinging" on our web of belief, leading to various adjustments in the latter, according to some principle of "minimal mutilation." The picture is very much one of a passive observer who responds to nature by engaging in belief adjustment but does not act on or change nature. The "action" component of pragmatism goes missing here—or rather the only kind of action contemplated is verbal behavior—"lo, a rabbit" as one bounds by. A satisfactory philosophy of science should reinstate the action element, which immediately suggests that we interact with the world causally, and do not just form beliefs about it.

2.4.9 Focus on Practice, Not Just Verbal Behavior and Belief

Reinforcing this passive observer picture is another focus that is characteristic of a number of pragmatists—again Quine and Rorty come to mind—and that is suggested by the observation under Section 2.4.8 above. This is an almost exclusive focus on language use and verbal behavior and (what are often taken to be their inner analogs) beliefs. (Among philosophers of science, this may be broadened to include "theories" construed as systems of beliefs.) But while language, the making of true statements, or more broadly the exhibition of models, theories etc. that are intended to convey information about nature are certainly important in science, a pragmatic philosophy of science should bear in mind, that science consists in a lot more than this—there are, again, activities that involve building, manipulating, synthesizing, and so on. A core notion for pragmatist philosophers of science should thus be the notion of a scientific *practice*, understood to encompass both verbal behavior, belief formation, and the construction of representations but also many other forms of behavior linked to choice and action.

2.4.10 Subject versus Object Naturalism

Another way of describing the interpretive/evaluative emphasis in Sections 2.4.1 and 2.4.2 above (what works and so on) is in terms of a focus on what practitioners are *doing* when they engage in various forms scientific practice, on what the *point* or *function* (or goal) of the practice is or the role or roles it is playing. I see this focus as closely connected to Huw Price's notion of *subject naturalism* and the contrast he draws between this and what he calls *object naturalism* (e.g., Price 2011).

Roughly speaking, the object naturalist tries to understand some body of discourse in terms of the objects the discourse is about or which serve to "ground" the discourse (and where these objects are construed in a way that is acceptable to naturalists—that is, as part of the world that science reveals) while subject naturalists instead focus on understanding what subjects do with the discourse, again in a naturalistic vein (so that disciplines relevant to such understanding may include, for example, anthropology and psychology). This frames the contrast in terms of two different ways of approaching *discourse* but, as already explained, I would favor broadening the focus (particularly in the case of subject naturalism) to include forms of practice that are not discursive, as well as those that are. In any case, the central feature of subject naturalism, as I understand it, is its emphasis on the importance the role of the *subject* or the *user* (typically a human being) who engages in various practices. Human beings have particular features that are relevant to this subject role: they have specific goals and interests, and various computational and manipulative abilities and limitations and these inform the science they produce. What we want to understand is what such subjects are aiming at accomplishing when they engage in various scientific practices, what the point or goal of those practices is and the extent to which they are successful. (This is an idea that I have particularly tried to emphasize in my own work on causation—see e.g., Woodward 2014, 2021a.) At least in many cases, these are very different concerns than the object naturalist concern with getting the ontology right or identifying objects in the world allegedly corresponding to or grounding these practices. As explained below, I see an impulse toward object naturalism in inappropriate domains as one source of ambitious metaphysics.

Pragmatists, I assume, will want to adopt some form of naturalism but as Price's discussion reminds us, there are different ways of being a naturalist about some body of discourse or practice. Often an important part of the most promising strategy for naturalists will be to try to understand what subjects are doing within a naturalistic framework, rather than just trying to "naturalize" the products (discourse, representations, etc.) they produce by providing an account of the "objects" they are about. For example, we should try to understand what subjects are attempting to do when they engage in causal reasoning (what their goals are) rather than just focusing on questions about what causation "is." Because of its focus on subjects and what they are doing, a subject naturalist treatment of science yields a picture according to which science, even physics, is not "the view from nowhere" (there is no such view) but always the view of concrete, situated inquirers. Thus, for example, limitations in what it is possible for humans to compute or calculate or manipulate should not be dismissed as irrelevant to the theories of science we construct or the concepts that we use to understand science. Indeed, a sensible methodology of science (which after all, within a pragmatist framework, is one that should be useful to scientists) is not possible without taking account of these limitations.

Finally (and again to anticipate) let me emphasize that this focus on the subject is (of course) not at all meant to imply that the world or nature should play no role in the account of science we seek to provide. On the contrary, as emphasized by Mark Wilson (2017) among others, subjects make use of and exploit various features of the natural world when they engage in scientific practices and predict and manipulate and so on. The world needs to cooperate (to provide support or enabling conditions) if these practices are to be successful. It is thus perfectly in order (and a very good question) to ask, for example: what is it about the world that enables us to fruitfully employ causal reasoning or particular kinds of causal analysis.[13] Relatedly, we can ask what information about the world is conveyed or captured by such claims. Similarly, if a certain computational procedure returns reliable results we can ask what features of the world support this procedure.[14] However, these roles for the world are not very well captured by philosophical treatments which make use of ideas about correspondence and isomorphic representation in trying to understand science—a subject to which I now turn.

2.4.11 Skepticism about (or Rejection of) Representation

Historically, many pragmatists have been skeptical about (or hostile to) "representation" as a useful concept for understanding how language and other sorts of activities such as scientific theorizing work. This attitude continues in the work of more contemporary pragmatists like Rorty and Price. My own view is that the appropriate target for this sort of skepticism is not representation per se but rather certain conceptions of what representation—particularly representation via models and theories in science—involves. These are conceptions that understand representation in terms of notions like mirroring, picturing, correspondence, isomorphism, and the like. I will call this conception *literalism*.

In what follows, my focus will not be on representation in general (I leave open the issue of how sentences of "the cats is on the mat" variety should be

[13] Again, one can think of such issues as part of metaphysics but if so, I claim that they only involve "minimal metaphysics."

[14] In this respect I fully agree with Ruetsche (this volume) when she writes:

"I don't take 'what theories represent' and 'what theories are for' to lie on opposite sides of a methodological fault line. Focus on one aspect needn't preclude focus on the other. After all, a theory's representational capacities could have *something* to do with its patterns of and aptness for use, and our aims in applying it could shape its representational capacities!"

More generally, attention to the role of the user of models and theories doesn't mean that we should neglect the capacity of the theory to provide information about nature; instead the two are complementary.

understood) but rather on certain very popular claims about how to understand the representational role played by (i) structures like theories and models in science and, along with this, (ii) the role of modal concepts like "cause" and "physical possibility." And as I have already signaled, I think that what we should reject is not the idea that successful models and theories represent but rather a particular way of thinking about representation in science—that this is always or even usually best understood in terms of metaphors like picturing, correspondence or (to use the currently fashionable term) the idea that a successful theory requires the existence of an isomorphism (or some notion linked to isomorphism—e.g., a partial isomorphism) between a successful theory or model and the system modeled. I think one can reject this idea, without rejecting the idea that theories, models, and other structures convey or allow us to extract *information* about what the world is like. So our guiding slogan should be something like: *information without (necessarily) isomorphism* or *truths about nature without literalism.*[15] In other words in rejecting representationalism, conceived in terms of picturing or isomorphism, we needn't embrace some form of idealism, anti-realism or views that regard scientific theories or models as mere "fictions." Nature and the world remain as central parts of the story and theories and models tell us about these; it is just that the relations between the two need not be what picturing metaphors suggest.

2.4.12 Against Eschatological or End-of-Days Science

A common way of approaching many issues in philosophy of science, especially among philosophers inclined to ambitious metaphysics, begins with the idea of a final or fully completed science—a theory of everything with nothing left out. (Presumably this will be largely or entirely a theory in which physics plays a fundamental role.) One then tries to understand features of contemporary incomplete science, including the "special" sciences,

[15] If this seems oxymoronic the idea is that an overly literal interpretation of what the theory says may not capture the truths it conveys about nature. As a simple illustration, a model in evolutionary biology may postulate a population with an infinite number of rabbits. It would be a mistake to interpret this as a claim about the cardinality of the rabbit population (the literal interpretation); its use is rather a way of claiming that certain other things are true of the population—e.g. that drift is unimportant and that mating is random or nearly so. These last claims will be straightforwardly true or false and the model will be appropriate if they are true or nearly so. We don't need to appeal to explanatory falsehoods or fictions to capture what is going on in such a case.

by reference to this theory. For example, Michael Strevens (2008) begins with the idea of a complete description of Causal Reality (as he calls it) of a sort that presumably would be described by some final physical theory (or perhaps the starting point is Causal Reality as it is in itself, apart from any description of it) and then attempts to understand current theorizing, in both physics and the special sciences, as the result of a process of "abstraction" (by which he basically means dropping details) from this ur-description. David Lewis begins with the notion of the full Humean mosaic—a record of what happens everywhere and at all times throughout the entire history of the universe—and then attempts to understand laws of nature in terms of the systemization of this information.

A pragmatist philosopher of science of the sort I am envisioning has a very different starting point: the goal is to understand "actually existing science" as carried out by human beings as they are actually are, given what they can know, calculate, and manipulate. Humans do not have access to all of Causal Reality, the full Humean mosaic, and similar sorts of information and yet somehow manage to do science, sometimes successfully. The goal of the pragmatist philosopher of science should be to understand how that is possible—and in the nature of the case, science as it will exist at the end of history, the full description of causal reality and so on cannot be part of this story.[16] Saying that current science reflects the processing limitations of human beings when dealing with the full Humean mosaic or that it approximates God's summary for beings with our limitations does not provide the right story about how our limitations come into the picture. There is no God to provide the summary and we do not "deal with"—either causally or otherwise— the full Humean mosaic which we then theorize about in ways that reflect our processing limitations.

Adoption of a pragmatic approach to philosophy of science suggests a major shift of focus away from "this end of days" picture. Philosophers of science instead should attempt to understand how scientists learn even in the presence of very partial information and limited computational possibilities—for example, how strategies that involve what Mark Wilson (2017) calls "physics avoidance" work, and how strategies like randomization work to enable the discovery of causal relationships even when a great deal about the detailed behavior of some system of interest is not known. Rather than asking what science would look like (what our concept of law

[16] Ideas about "reference magnetism" might be understood as claiming otherwise but they come with metaphysical commitments that will not be attractive to pragmatists.

of nature etc. would be) if we knew everything, we should take seriously the idea that we don't know everything and consider how this affects the content of our science.

2.4.13 More on Metaphysics

I distinguished earlier between minimal and more ambitious metaphysics and suggested that pragmatist philosophy of science requires only the former.[17] Part of my motivation for this claim is that it appears that even if one were to have definitive answers to the fundamental questions addressed by ambitious metaphysics these would tell us very little about the methodological/interpretive issues that are the province of the pragmatic philosopher of science. Indeed, ambitious metaphysics, at least as currently practiced, seems largely uninterested in these issues: To take just one example, to the best of my knowledge neither metaphysicians who claim that causal claims are grounded in laws of nature conceived in accordance with the best systems analysis nor the metaphysicians who favor non-Humean treatments of causation make any serious attempt to connect their views with the methodological issues about causal reasoning and inference(issues of the sort that are discussed in e.g., Pearl (2000), Spirtes, Glymour, and Scheines (2000), Morgan and Winship (2015), and Woodward (2014) that are natural topics for pragmatist philosophers of science. Moreover, there are systematic reasons for this lack of connection: as I have noted, the willingness of most metaphysicians to abstract away from what they regard as merely practical barriers to what we can know or calculate and other human limitations virtually guarantees that many of their inquiries will be of limited relevance to the descriptive and methodological questions that interest the pragmatic philosophers of science.

For these reasons pragmatist philosophers of science should feel free to insist that, given their goals, they don't have to do ambitious metaphysics and that the projects they wish to pursue are important and valuable in their own right. When criticized by metaphysicians for failing to articulate proper

[17] This proposal cross cuts some other possibilities for engaging with metaphysics. For example, John Dupré (2012) advocates the development of a distinctively pragmatist metaphysics, organized around notions of process, at least for biology. Depending on the details, this might be a kind of minimal metaphysics, centered on general empirical claims about the nature of biological phenomena and how these fit with practices of biological investigation and theorizing. Alternatively it might involve far more ambitious forms of metaphysics, as in the process philosophies of Whitehead and Hartshorne. I take Dupré to have the former in mind.

metaphysical underpinnings for science, pragmatists should push back by demanding to be shown why, given their goals, such underpinnings are necessary.

2.4.14 Reductionism

A number of philosophers of science, again especially those who are inclined to ambitious metaphysics, are attracted to one or another variety of reductionism about various concepts employed in science.[18] A common focus of such reductive programs is causal and modal notions, with so-called Humean reductivist projects being very prominent among these. The questions a pragmatist philosopher of science should ask about such projects are these: (i) First, given the goals of the pragmatic philosopher of science as described above, how if at all would successful completion of the reductivist project contribute to these goals? If, as I suspect, carrying out the interpretive/evaluative projects described above often does not require carrying out a reduction, pragmatist philosophers should feel free not to provide it. (ii) More generally, the pragmatic philosopher should ask the ambitious metaphysician just what the point or purpose of the reductions she wishes to provide are. What does a reduction give us (beyond the reduction itself and the satisfaction of a metaphysical itch to reduce)? In the past, it was common to argue that in the absence of a reduction, we did not really understand causal and modal locutions or that we lacked a story about how we could acquire knowledge about them. But these semantic and epistemological claims are no longer plausible. Is there some other reason why reductions are valuable? Or to put the question a bit differently: suppose we have the pragmatist goals described above in connection with causal reasoning—understanding the use of causal notions, how we reason with them, infer to causal claims and so on. What reason is there to think that failure to carry out a reduction would impede those goals?

2.5 More on Subject and Object Naturalism

Having set out these themes, I turn now to an attempt to develop and motivate them in a bit more detail, beginning with the distinction, due to Price

[18] My target here is reduction of one set of concepts like cause, probability, etc. to another, not the reduction of one scientific theory to another.

and alluded to earlier, between object and subject naturalism. Price characterizes object naturalism, at one point, as the view that "all there *is* is the world studied by science" and, as an epistemological doctrine, as "the view that all genuine knowledge is scientific knowledge" (2011, 5). He contrasts this with subject naturalism according to which "philosophy needs to begin with what science tells us *about ourselves*" (p. 5). As Price explains, when applied to some bit of discourse, the subject naturalists asks what we humans are *doing* with the discourse, what its goal or point or function is, and tries to understand this (and the subject engaged in this activity) naturalistically.

Since I'm not sure that I understand either the claim that what is studied by science is all there is or its denial, I'm going to adopt a slightly different characterization of object naturalism from the one quoted above, but one that I think remains in the spirit of Price's usage. This construal takes object naturalism to involve a thesis about (or project motivated by assumptions about) *representation* (that it is to be understood in terms of notions like mirroring or picturing or correspondence or literal truth) and an accompanying thesis about what is required for a representation to be true or fully accurate: roughly speaking, the representation claims that certain *objects* exist—the objects that the representation would mirror or picture if it were construed literally— and the representation is true or accurate to the extent that this is so.[19] In other words, the objects that are required to exist can be read off in a straightforward literal way from the representation itself. (I will understand the notion of "object" here very broadly so that it includes properties or relations, but one of the rules of the game is that these are understood as having thing-like characteristics, so that one can raise questions about their "ontological" status.) For example, claims about what is possible

[19] The metaphysics literature on laws of nature provides one illustration of this tendency to interpret scientific claims as claims about the existence of objects, even when this seems inappropriate. In a recent survey, Bhogal (2020) describes laws of nature as "things like the Schrodinger equation" and suggests that many anti-Humeans treat laws as "separate governing entities." Bhogal's sympathies are with Humeans but (apparently) not because he thinks there is anything wrong with thinking of laws as things or entities—he just thinks the things with which laws are to be identified are patterns or regularities in Humean mosaic. But even putting aside this elision between the Schrodinger equation, which is a mathematical structure and what in the world it is answerable to, it seems bizarre, whether or not one is a Humean, to think of laws as things or objects, with the accompanying thought (if one is a non-Humean) that these might stand in "governing" relations with more ordinary objects. Laws describe how the behavior of some objects depends on the states of others, but laws are not entities that make things happen in the way that electrons do. Pragmatists need another way of interpreting laws besides in this object oriented way.

are understood as claims about the existence of "possibilities" or "possible worlds" and so on, the ontological status of which then become matters of concern. If one can find such objects for all of the claims in the discourse and these objects are naturalistically acceptable, then these claims are taken to be true (legitimate, etc.). On the other hand, suppose the attempt to provide such a construal runs into difficulties, as one might think that it does with respect to, say, moral claims or claims involving physical modality, where one faces puzzles about the objects such claims are about or the things that serve as "truth-makers" for them. Then one faces what Price calls a *placement* problem, which might be resolved in a number of different ways—e.g., by expanding one's view of what is a naturalistically acceptable object or, alternatively, by postulating objects that are acknowledged not to be naturalistically acceptable in order to secure the truth-aptness of the claims in question. Yet another alternative is to reconstrue the claims as not about objects at all and hence (on the view of representation we are considering) as not (at least literally) the sorts of things that can be true or false. (Perhaps the claims should instead be understood "expressively" or as "useful fictions.") In any case, the focus is on *existence* claims and on what objects in the world correspond to or make true the claims or representation in question.[20] In other words, in contrast to subject naturalism, the focus is ontological, even if the conclusion is that the appropriate objects to make the representation true do not exist.

This suggests how a kind of literalism about representation combined with the kinds of concerns that underlie object naturalism tend to lead to ambitious forms of metaphysics when applied to discourse and practices whose corresponding objects are not obvious. Either one is led to postulate objects whose existence seems puzzling or controversial to serve as such correspondents and hence to worries about the ontological status of these or one is led to controversial claims about the discourse in question being fully reducible to claims about objects that are naturalistically acceptable. Alternatively, one reinterprets the discourse as not truth-apt at all.

Using one of Price's running examples, consider moral claims and discourse as an illustration of these possibilities. Presented with the contention that stealing is wrong, a philosopher with an orientation toward object

[20] Discussions of whether numbers "exist" (where this involves asking what sorts of objects numbers are) and whether when mathematics is used in science, this requires an "ontological commitment" to the existence of numbers are an extreme example of the tendency to try to interpret a discourse in terms of the "objects" it is about.

naturalism would consider, first, whether there is some property or thing corresponding to "wrongness" that is instantiated by various acts of stealing. Candidates for such a property that might be argued to be naturalistically acceptable may include, say, "not conducive to maximal preference satisfaction or welfare maximization."[21] Alternatively, the investigator might want to consider giving up on naturalism and postulating some non-natural property of "wrongness" to serve as a truth maker for claims about the wrongfulness of stealing. If one thinks that plausible candidates for such properties do not exist, another alternative is to construe the claim as not a candidate for literal truth of falsity at all, and to re-interpret it as, say, (merely) an expression of disapproval regarding stealing.

The subject naturalist takes a very different approach: rather than focusing (just) on what the objects and properties might be that are the worldly correspondents to moral claims, the subject naturalist focuses instead on what speakers are doing when they make such claims, or when they engage in practices associated with moral assessment more generally. The subject naturalist asks questions like "what is the point (or points) or goals or function of these practices?," "How (if at all) does labeling actions as wrong contribute to these?" In the case of moral claims, for example, such goals might have to do with coordination with others about rules governing social interaction in order to advance certain commonly held social goals, the advancement of proposals that others adopt and act on certain rules (where this might involve arguments that the rules are feasible and motivating for others) and so on.[22] On a descriptive level this is an inquiry that can and is pursued by anthropologists, sociologists, economists, and others. And of course there are evaluative projects along these lines that might also be pursued: one formulates various social goals and then asks whether various moral practices or judgments are conducive or not to realizing them.

As already suggested, as I will understand subject naturalism, it *need not* involve the view that the claims (or representations or practices or whatever) in question are the sorts of things that are not true or false or that fail to have import for how matters stand in the world. That is, one does not have to be some kind of subjectivist or relativist about the practices in question to engage in a subject naturalist approach to them. What is distinctive

[21] In fact, these are dubious candidates for *naturalistically* acceptable properties but let's put this consideration aside.

[22] At this level of generality such goals may seem anodyne and uninformative. They acquire more content when combined with assumptions about feasibility etc.

about subject naturalism is that the *focus* is different from the focus of object naturalism, in the ways that I have tried to characterize—the focus is, as Price says at one point, "anthropological"[23] (to which I would add, "or, alternatively, evaluative") rather than "ontological." However, it is consistent with this focus that the best way of understanding some practices (including practices of scientific theorizing) is to understand them as in the business of providing information about how matters stand in the world.

While a subject naturalist treatment of a practice or discourse thus does not require construing it as non-truth apt, subject naturalist approaches can open up, in a way that I will illustrate, the possibility that literalist construals in terms of mirroring and correspondence may not be the best way to understand those practices. So a subject naturalist understanding of some practice may help to undermine certain literalist understandings of the practice (and accompanying claims about the objects to which the practice is committed), although in doing so it need not also undermine interpretations of the practice as providing information about nature. This is because a better understanding of what subjects are doing when they engage in some practice can lead us to rethink what, so to speak, the worldly correlates of the practice are and the structures in nature that account for the success of the practice.[24]

It is striking (and further evidence for the importance of the distinction between subject and object naturalism) that the accounts provided by object naturalists are often rather unilluminating regarding the issues that interest subject naturalists—they just don't address the subject naturalist's concerns. Relatedly, it seems entirely possible for subject naturalists to pursue their concerns without committing themselves on many of the issues on which object naturalists focus. This is illustrated by the example of morality. Subject naturalist investigations of this topic flourish in many different disciplines but for the most part these successfully avoid taking up issues having to do with the ontological status of "wrongfulness" and the like. Conversely, if we were told by some oracle that, say, objective wrongfulness exists as a non-natural property or, alternatively, that it can be identified

[23] I take it that anthropological in this context means something like "descriptive/interpretive." Historical or psychological investigations can thus fall under this rubric.

[24] As Price remarks, there is a tendency to suppose (assuming one wants to be some kind of naturalist) that object naturalist concerns are prior to and more fundamental than subject naturalist concerns. I agree with Price, however, that matters are often the other way around, in part because without some (subject-naturalist) account of what one is doing with some bit of discourse (or theory or model), one does not know whether or in what respects an object naturalist construal is appropriate.

with some complicated natural property, this would tell us very little about what subjects do when they engage in various moral practices, make moral judgments and so on. Nor need it help in any straightforward way with moral evaluation. Indeed, should they be forthcoming about such matters, it seems to me that most of those interested in the ontology or metaphysics of morals will agree that they are not interested in such descriptive or evaluative questions, and that they instead regard these as mainly the concern of other disciplines such as psychology and anthropology or perhaps, in the case of evaluation, social policy analysis.

Returning now to a topic of more direct interest to philosophers of science, I think that a similar point holds for investigations into causation and causal reasoning. Here too I see two quite distinct foci of interest, corresponding to object and subject naturalist approaches. The object naturalist wants to know what in the world corresponds to the causal relation (or to the variety of such relations, but often for the object naturalist the assumption is that there is just one). What is the "nature" of causation or the "grounds" or "truth makers" for causal claims or the correct ontology/ambitious metaphysics for causation? By contrast, subject naturalists have the usual mixture of descriptive and evaluative concerns—at the descriptive level, they want to know how various subjects reason about causal relationships, how and on the basis of what sorts of evidence subjects infer to causal relationships, what sorts of goals inform causal thinking (why we care about discovering causal relationships rather than resting content with "mere correlations"), what distinctions subjects make among different sorts of causal relationships and so on. These are issues that are addressed by psychologists interested in causal cognition but also by a number of philosophers—for example, by Danks, Glymour, Hitchcock, and Wilson, among others. On the evaluative level we have a vast literature on the methodology of causal inference and reasoning. Once again, I think there is a disconnect between these concerns and more purely ontological ones. Even if some oracle were to tell us that the correct ontology for causation is such and such, it is not clear that this would provide much by way of answers to the sorts of descriptive/methodological questions described above. Conversely, it would seem that researchers outside of philosophy who pursue such questions do so without much attention to object naturalist/metaphysical questions about causation and it is not at all obvious that they are wrong to do so (cf. Woodward 2014).

I assume that you can see where this is going. I want to recommend the subject naturalist attitude or approach not just in connection with understanding the role of causation in science but also more generally in philosophy of science (although again not to the exclusion of minimal metaphysics). An approach in which subject naturalism figures prominently fits naturally with and supports both the descriptive/interpretative and the evaluative concerns of philosophers of science that are a natural expression of a broadly pragmatist approach to science.

2.6 Means/Ends Reasoning

Let me next turn to some remarks about goals and means ends reasoning. There are very influential strands of pragmatism, associated with Quine and Rorty among others, that aim at collapsing or undermining distinctions. Where others see differences, these pragmatists see continuity, sameness, and seamless wholes: theories confront evidence holistically in a way that prevents our regarding different components of those theories as less or more well supported by evidence; science itself is claimed to be continuous with and not to differ in principle from pure mathematics and perhaps even from metaphysics. A common device among these pragmatists for motivating such views is to invoke the idea that it is appropriate to think of all elements of science as animated by a single overriding undifferentiated goal—that of finding what "works" or "is useful" or "fruitful." The assumption is that anything that contributes to this generic goal has a similar status and that further distinctions are unnecessary and even invidious. Thus, for example, just as we allegedly "posit" the existence of electrons because it is "useful" to do so, so also we posit the existence of sets and natural numbers because this too is useful.

Notice that this idea that there is one overriding goal of usefulness is distinct from the characterization of the pragmatic approach adopted in Section 2.4 above. Section 2.4 says that science is concerned with such useful goals as prediction and control and that various aspects of scientific practice are to be understood and evaluated in terms of their conduciveness to these goals. Nothing is said about there being a single undifferentiated goal of usefulness.

In contrast to this notion of a single goal of usefulness, the sort of pragmatism that I favor takes seriously the idea that human beings have a number of different goals, both within scientific investigation and elsewhere. These goals carry with them different standards of adequacy and success. For example, even at a very general level, prediction is a different goal than causal analysis and it is unhelpful to subsume both of them under the single goal of discovering relationships that are "useful" or that "work," even though it is of course true that both prediction and causal analysis can be useful in their different ways. Pragmatists should be sensitive to these differences rather than collapsing them into a single ur-goal. In addition, on the approach I favor, a goal like prediction is usually understood as a matter of making *true or true enough or accurate* predictions—there is no suggestion that notions like truth or accuracy themselves can be understood in terms of some undifferentiated notion of usefulness. There is also no suggestion that we are entitled to "posit" claims as true just because doing so would be useful to us in some general way—I take this to be an endorsement of wishful thinking and antithetical to science.

Thus, the version of pragmatism I favor presupposes the availability of some notion of truth.(I take this to include some notion of being nearly true or close enough to true as to make no difference, as when one says that Newtonian gravitational theory is true or valid for large objects in weak gravitational fields). It does not attempt to analyze truth in terms of usefulness or warranted belief or anything similar. Instead, the assumption is that we need a notion of truth to specify in a non-trivial way which means in fact lead to goals, which reasoning strategies lead to correct conclusions and so on.

I have already mentioned, as among the goals associated with science, prediction, explanation, and causal analysis, classification, building and making. These goals can in turn be individuated more finely in many different ways. For example, there are many different kinds of control-related questions to which investigators seek answers, and many different kinds of predictions, distinguished both by their targets and the information on which they rely.[25] Moreover, investigators have choices (i.e., they may have different goals) about which behaviors of which systems they wish to predict, explain, and so on. And of course, realization of these goals will require realization of more specific subgoals—e.g., successful prediction may require accurate measurement of some quantity of interest, construction of instruments for doing this and so on. Nonetheless the goals themselves are distinct and, as I said above, can have different success conditions. For

[25] For example, in causal analysis, the notion of a total effect corresponds to a different manipulation- related question than the notion of an effect along a causal path—see Woodward (2003). Finding a function that maximizes accurate predictions for random draws from an i.i.d distribution on the basis of previous draws from the same distribution is a very different problem than predicting the future positions of the planets and so on.

example, models or choices of variables that are useful for certain predictive purposes may not be optimal for purposes of explanation and causal analysis, as recent work by Danks (2015) and numerous other examples show.

Even when we are interested in explaining aspects of the behavior of a single system, different models and strategies will often be appropriate, depending on which aspect of the system or the level or scale of its behavior we are aiming to explain. For example, a circuit-level model like the Hodgkin–Huxley model of a neuron, which seeks to explain the generation of an action potential, has a different target explanandum (or set of such explananda) than a molecular level model that aims at the explanation of the opening and closing of the ion channels that are responsible for the ionic currents that are at work when the action potential is generated (Herz et al. 2006). Pragmatists will find it reasonable to look for an understanding of explanation according to which one can explain one of these explananda (satisfy one goal) without explaining the others—see Section 4.10.

One corollary of this is that, corresponding to this plurality of goals, pragmatists should expect to find in science a plurality of theories or models, even in connection with the same kind of system, depending on what goals inquirers have in connection with that system—what scale of behavior they want to predict or explain and so on. At the same time, investigators' goals will often require establishing connections or relationships among different models. For example, for certain predictive purposes it may be adequate to model a material as a homogeneous continuum. But for other predictive purposes (e.g., in modeling cracks and fractures) it may be best to continue to use a continuum like model but feed it information from lower scales—as it is sometimes put, finding ways of allowing lower scale models to "talk to" more upper scale models and vice-versa.[26] We should expect to find not just a plurality of models but attempt to integrate and connect these—a picture that fits very well with Sandra Mitchell's integrative pluralism (e.g., 2002); see Section 4.10 for additional discussion.

Let me also add in this connection that my view is that it is only if we acknowledge a plurality of goals, individuated in a somewhat fine-grained way, that we can carry out the evaluative and interpretive tasks described above in a way that is interesting and illuminating. If we think in terms of specific goals like estimating the value of some parameter from a body of statistical data or inferring causal relationships from observational data

[26] See Batterman (2013).

meeting certain conditions, we can describe and evaluate specific procedures for achieving these goals and in this way provide useful guidelines for the conduct of inquiry. If we insist on framing discussion around highly generic and imprecise goals holding for all inquiry, whether this be "usefulness" or "accommodating observational evidence with minimal changes in existing theory" or something similar this is unlikely to generate helpful guidelines.[27]

In my list of possible goals above, there are conspicuous omissions: For example, I did not include the goal or goals of conveying the full and literal truth about nature "at the fundamental level," or of "providing a single completely accurate representation of what nature is like in every respect." I have several justifications for this omission. First, I am skeptical that, as an empirical matter, this goal plays a central role in science. Second, I doubt that this and related goals are attainable or even fully intelligible. Indeed, I think that such alleged goals are part and parcel of the literalist idea that scientific theories and models aspire to some sort of mirroring or isomorphism relation with nature and, as explained above, I favor rejecting this whole package. Third, even if the goal of full and literal truth is among the goals of science, I think it obvious that it is not the only such goal. And more importantly, as we shall see, this literalist goal is often in tension with the other goals described above. For example, it is often not true that the model or theory which is most representationally detailed or accurate provides the most adequate explanations. So even if we think isomorphic representation and full and literal truth are goals, these are often overridden by other goals. Note also that to the extent that the project of reading off the correct metaphysics from our best scientific theories depends on the idea that the only goal of scientific inquiry is something like isomorphic representation of the fundamental nature of reality, that project will be problematic if science reflects other goals as well that are in tension with the isomorphism goal.

[27] Lots of claims in contemporary philosophy of science are framed at a level of generality and vagueness that make any sort of serious means/ends analysis impossible—for example, claims about the virtues of "inference to the best explanation." In standard treatments, the notion of a "best explanation" is left unspecified and there is also no analysis of why this "method" of inference should be expected to lead to truth or anything similar. In my view, pragmatist philosophers of science should avoid claims and concepts of this sort. A similar complaint holds for many appeals to "simplicity" in philosophy of science (see below).

2.7 Means/Ends Methodology Illustrated

Let me turn next to some additional explication of the means/ends picture of science and its methodology gestured at earlier.[28] Recall the basic picture: scientists have certain ends or goals and many aspects of science—particular choices of methods and choices about modeling and theorizing should be understood as means to attaining those goals and evaluated accordingly. The category of "means" is understood very broadly—it includes strategies for experimental design, procedures for causal inference from various sorts of data, choices of variables in model construction, testing procedures of various kinds and much more. A simple illustration is provided by the treatments of estimators in classical (frequentist) statistics. Suppose we wish to estimate the value of some quantity m on the basis of noisy measurements resulting in statistical data d bearing appropriately on m. Such an estimator will be a random variable m^* that is a function of the data d. One proceeds by adopting criteria for what it is for an estimator to be "good." For example one such (widely used) criterion is that the estimator be unbiased in the sense that its expectation value $E(m^*)$ is equal to m – i.e., to the true value of the quantity being estimated. Another criterion is that the estimator should be chosen so that, among such unbiased estimators, its variance is as small as possible. Given these criteria and certain other assumptions, such as the assumption that the estimator must satisfy a linearity requirement, one can then prove mathematically that the best estimator for m must take a certain specific form.

What I want to draw attention to here is not the details of this idea but rather the general form that this justification takes: one specifies a goal—a good estimate for m, subsidiary goals that specify what counts as a good estimate in this context—and then shows that a certain choice of an estimating function is an effective means to this goal, thereby providing a means/end justification for this estimating procedure. This is exactly the sort of justification that a pragmatist should like and indeed in my view it is no accident that one of the founders of philosophical pragmatism, Peirce,

[28] Many other examples of means/ends reasoning in addition to those discussed below can be found in statistics, machine learning, and formal learning theory. Additional philosophical examples include Sober on black box inference, and Glymour (2001) on inference from lesion data in cognitive neuroscience. Let me add that although in this section I focus on means/ends reasoning in which the goal is truth or good error probabilities of truth and falsity, there can of course be other goals besides truth.

was an early source for the use of such frequentist or error-characteristic justifications in statistics.

2.7.1 Causal Claims

In the example just described the chosen goal is rather local and the demonstration that the means are optimal is purely a matter of mathematics. But, as already intimated, I think that the general idea can be applied much more widely and used to illuminate and evaluate many other concepts and reasoning strategies used in science. (As noted above, in the more general case, means/end analysis may involve some mixture of empirical evidence, heuristic argument, strategies for calibration, and simulation as well.) For example, consider an interventionist treatment of causation according to which causal claims are understood about as claims about what the results of hypothetical experiments would be. Take the discovery of the truth about such claims as the goal at which we are aiming. One can then raise questions about the kinds of strategies, evidence, and methods that are likely to be effective in realizing this goal. One thing that is immediately suggested is that when presented with a causal claim, it pays to try to clarify or disambiguate it by specifying as exactly possible just which hypothetical experiment is intended (as capturing the content of the causal claim), since what looks like the "same" causal claim can often be associated with different possible hypothetical experiments. This is common methodological advice in at least some portions of the literature on causal reasoning and inference.

Having clarified which hypothetical experiment is intended, one can then assess the extent to which different bodies of evidence and inference procedures are effective means to this goal. Obviously one very simple procedure is to actually *do* the relevant hypothetical experiment, since this will tell you what would happen if you were to do it. But even if one can't (or doesn't) perform the relevant hypothetical experiment conceptualizing causal claims in the manner described can be very useful: it suggests a program of evaluating inference methods in terms of what they can tell us about the outcomes of hypothetical experiments without actually doing the experiments. And in fact, certain procedures such as the use of instrumental variables and regression discontinuity designs can, under the right conditions, be demonstrated to furnish reliable answers to such questions—indeed this is a standard way of justifying the use of such procedures. In the case of instrumental variables, for example, one looks for relationships among the observed or measured variables that mimic the effects of interventions, so that one can use this information to identify what the result of an intervention on the candidate cause variable would be without actually performing the intervention in question.[29] Again, this is means/ends reasoning: if you want to predict the outcome of a hypothetical experiment without actually doing the experiment, you should proceed in such and such a way. That is, if you follow the appropriate procedures and the relevant assumptions are satisfied, it follows, as a matter of mathematics, that you will correctly predict the results of a hypothetical experiment.

2.7.2 Invariance

As another illustration, also in the context of causal reasoning, consider the role played by notions of invariance. A relatively invariant causal relationship is one that is relatively stable or insensitive in the sense that it continues to hold across changes in other variables. It should be obvious why finding relatively invariant relationships is worthwhile goal—among other considerations, such relationships are more generalizable to new circumstance and afford wider opportunities for manipulation and control. Given this goal, various strategies and practices can then be understood in terms of their conduciveness to this goal. As an illustration consider our interest in discovering "mechanisms." Given an overall relation of dependence between some input *I* and output variable *0*, the individual relationships between mediating or intermediate variables between *I* and *O* will often be more invariant than the overall $I \rightarrow O$ relationship. Guided by an interest in finding relevantly invariant relationships we look for such links; we decompose the black box $I \rightarrow O$ relationship into intervening steps or each of which, taken individually involves relationships that are more invariant that the overall $I \rightarrow O$ relationship. These comprise the mechanism linking *I* to

[29] Suppose that the data generating process is represented by $Y = bX + U$, with *X* and *Y* the candidates for cause and effect variables, and *U* an error term that is *correlated* with *X*. As is well known, when *X* and *U* are correlated, we cannot use the simplest possible estimator for *b*, involving ordinary least squares. We can, however, reliably estimate *b* if we can find an instrumental variable *Z* for *X* with respect to *Y*. There are various ways of characterizing such an instrument. One is that *Z* is an instrumental variable that (i) is associated with *X*, (ii) is independent of *U*, and (iii) is independent of *Y* given *X* and *U*. If *Z* meets the above conditions, then $b^* = cov\ (Y,Z)/cov\ (X,Z)$ is an estimator for *b*, interpreted as the *causal effect* of *X* on *Y*. Within an interventionist framework this interpretation is justified because (or to the extent that) Z functions as an intervention on *X* with respect to *Y*.

O. Notice the form that this takes: we don't just posit an interest in finding mechanisms as primitive or invoke it because, as a descriptive matter, scientists seem to be interested in looking for mechanisms. Instead, we try to explain why this is interest is reasonable or intelligible as a means to other more general goals such as the goal of finding invariant relationships which is in turn seen as contributing to the even more general goals described above. One advantage of such an approach is that it also illuminates why, in some circumstance or with respect to systems with certain structures, the search for the specific sorts of invariant relationships associated with mechanisms may not be such a good strategy—it may not serve more general goals of the sort just described very well. (See Woodward 2013.)

Yet another illustration of the same basic framework is provided by recent work by Patricia Cheng and collaborators (e.g., Carroll and Cheng 2010). She shows that in some circumstances (different from those discussed immediately above) a concern with finding invariant relationships will justify or motivate the introduction of latent or unobserved variables, while in other ostensibly similar circumstances, it will not. This claim about when it is appropriate to introduce latent variables is a normative claim, again in the means/end hypothetical imperative sense described above, but Cheng also shows that subjects in her psychological experiments behave in accordance with these norms. It and similar investigations have the potential to cast light on the issue of when investigators, whether lay or scientific, introduce terms or entities that go beyond what is observed and when it is rational or justified to do so.

2.7.3 Variable Choice

Next consider issues having to do with vocabulary or variable choice in science. When discussed by philosophers of science (or metaphysicians) the guiding thought often is that certain terms or predicates or properties are "natural" and others are not and that good theories are (and should be) formulated in terms of the former rather than the latter. The natural predicates in turn are identified on an intuitive, example-driven basis, often relying on syntactic or else broadly metaphysical considerations—"grue" and "overly disjunctive" predicates are non-natural and bad, "purely qualitative" predicates are more natural. A striking feature of most of this discussion is that the good and bad predicates are not evaluated as such with respect to the goals or purposes they are being used to achieve—presumably because the "naturalness" of such predicates is thought to depend on metaphysical considerations that are independent of human goals and purposes.

An illustration of this is provided in Sider (2011), who imagines a universe consisting entirely of fluid that is divided by a plane into a half that is uniformly red and a half that is uniformly blue. He supposes that observers encountering this universe divide it instead by means of predicates that cross-cut this color division. Sider contends that in doing so the observers are making a "mistake" and that they are carving up the world incorrectly. In contrast to Sider, pragmatists will think that the question of which classification or choice of variables is "correct" cannot be answered independently of what the goal or purpose is for which those variables are to be used and how in turn these goals interact with the empirical circumstances in which the variables are used. In Sider's example the role of these considerations is suppressed because the imagined universe has no additional differentiating properties besides those described—it is left completely unclear what the observers might be trying to *do* with the distinction they introduce.[30] Consider instead a much more realistic example in which one's goal is to formulate a distinction or classificatory variable that enables prediction of the values of certain other variables. It is entirely possible that, for this purpose, the "unnatural" distinction described by Sider that cross-cuts a classification based on color may be better (in the sense of yielding more accurate predictions) than the allegedly more natural distinction that he favors.[31]

As I have suggested, a pragmatist philosopher of science will approach the problem of variable or vocabulary choice in a very different way from Sider, asking instead which such choices will best further various ends of inquiry, so that the choices are evaluated in a goal-relative way. Thus, rather than assuming at the outset on the basis of syntactical or other considerations that disjunctive predicates are defective, pragmatists will ask whether there is any means/end argument that shows that such variables or predicates contribute less well to our goals than alternatives. The introduction of new terms and variables, however intuitively unnatural or non-simple, will be regarded as justified when they allow for the formulation of theories etc. that best advance goals of inquiry. Moreover, pragmatists will recognize that because of the plurality of goals in science, it is entirely possible that the

[30] It might be thought that this question has an obvious answer—the observers are just trying to "describe" the situation they see. Again, I deny that this notion of pure description, independently of more specific goals and purposes, has any application to science.

[31] After all, as a matter of empirical fact classifications based on color are not very useful for many predictive and explanatory purposes, however "natural" they may seem.

variables or vocabulary that work best for purposes of one set of goals (e.g., prediction) may not work best for other goals (e.g., causal explanation). And of course different variables may be appropriate for causal explanation even in connection with the same kind of system depending on what one is interested in explaining.

A very nice example of this general strategy in connection with variable choice is provided by Hitchcock (2012). Hitchcock takes the "metaphysical" question of whether the time at which an event occurs is "essential" to it and transposes this into a question about variable choice. For purposes of causal analysis should we represent events of the same kind occurring at different times (e.g., a fire occurring in May or in June) as different possible values of the same variable or as values of different variables? As Hitchcock shows, the first choice leads to a causal analysis that is obviously defective in the sense that it does not adequately capture the causal possibilities and is a misrepresentation of causal structure, assuming that one takes causal claims to have to do with the outcomes of interventions—for example, a causal cycle is represented as present in a situation in which it is not. More generally, the resulting structure does not accurately represent the full range of manipulability relations or interventionist counterfactuals which we assume to be true in the situation as described. The second choice has none of these defects and thus better serves our purposes.

In a recent paper (2016), I have attempted to further explore how the goal of discovering causal relationships in the interventionist sense, where these relationships also have such additional desirable features such as relative invariance, can guide variable choice. My idea is that some choices of variables make possible the formulation of relationships that are more invariant, and have other features that are desirable from the point of view of causal inference than others. Again, this project is both descriptive/interpretive and normative/evaluative: the idea is both to explain features of the variable choices that scientists make in terms of the goals they are trying to achieve *and* to assess whether and to what extent these choices are well adapted to these goals.

2.7.4 Simplicity

Many of the examples I have discussed so far have to do with causal inference and explanation but, as I said above, it is possible to adopt a similar approach to other aspects of scientific practice. Consider the role of "simplicity" in science. One possible approach is to observe that scientists sometimes appeal to simplicity considerations in theory choice and to take this as primitive fact, perhaps making a preference of simplicity constitutive of good science (or constitutive of some important concept in science, such as that of a law of nature, as in the Best Systems Analysis of this concept), but without trying to be more precise either about what simplicity involves or why it is important or valuable. An alternative approach, which is "pragmatist" in the spirit I have been trying to articulate, instead asks whether and how simplicity conceived in this or that specific way might be conducive or not to various additional goals—in other words, it tries to justify appeals to simplicity in a means/ends framework. In order to carry out this project, one has to be precise about what is meant by "simplicity" and about the goal to which it is taken to contribute.

One paradigm of this sort of approach is the Akaike framework (e.g., Akaike 1974) for assessing predictive accuracy. Within this framework, one particular way of thinking about simplicity, connected (roughly) to trading off the number of free parameters in one's model or hypothesis against a measure of goodness of fit on already observed data can be shown to be an effective means of maximizing predictive accuracy on new data, provided certain additional (and rather restrictive) assumptions are satisfied. In effect, a particular conception of simplicity connected to number of free parameters is understood as a corrective to over-fitting known data. Again, my point here is not to argue in favor of this particular set of ideas (which to repeat, apply only in a rather restricted framework) but rather to provide another illustration of "pragmatic" means/end justification and analysis.

Another sort of means/ends justification of appeals to simplicity is provided by formal learning theory, as developed in a series of papers by Kevin Kelly and co-authors (e.g., Kelly 2007). These papers show, roughly, that a learning procedure that conjectures the simplest hypothesis (in a specified sense) will converge to the truth (on a hypothesis space that contains the true hypothesis) and, considering all possible data streams, will in the worst case converge to the truth with the minimal number of retractions. Convergence to the truth with the least number of retractions in the worst case is thus the goal in terms of which adoption of this particular understanding of simplicity is justified.

There are of course other ways of thinking about simplicity and its role in science in addition to the notions and justifications just described. One alternative conceptualization of simplicity is in terms of ease or economy of use, as when it is said that use of a representation in spherical rather than

Cartesian coordinates makes solution of a certain class of problems easier. There is an obvious means/ends justification for a concern with this sort of simplicity but we need to be clear about just what such a justification can deliver. In particular, choosing a representation on the basis of this sort of justification does not address the problem that the Akiake framework attempts to address. On this ease of use conception of simplicity, the same hypothesis can have both a simple and a more complex formulation, both of which (since they represent the same hypothesis) will be equally reliable or non-reliable in predicting new data. In general, showing that a hypothesis is simple in this ease of use sense seems to have no implications for whether it possesses such other desirable features as truth or evidential support.

A frustrating feature of many appeals to "simplicity" in current philosophy of science (and metaphysics) is that little attempt is made to distinguish among these different notions of simplicity and the alternative justifications that go along with them. Consider, for the example, the role of simplicity in best-systems accounts of laws of nature. Is this to be thought of just as a matter of ease or convenience of use, along the model of spherical versus Cartesian coordinates? If so, axiomatizations of the Humean mosaic which differ in "simplicity" will apparently just amount to alternative representations of the same set of facts—representations that don't really differ in what they claim about the world, but only ease of use, in the way that representations of the same curve in terms of Cartesian and spherical coordinates do. On this view of the matter, it seems to follow, within this framework, that different claims about what the laws are will differ only in the way that representations of the same hypothesis that are more or less easy to use differ. Alternatively, if some more substantive conception of simplicity is intended by advocates of the BSA, this conception needs to be spelled out and some account needs to be provided for why it leads to the successful identification of laws of nature.

2.7.5 Physics Avoidance

As a final illustration, consider Mark Wilson's notion of "physics avoidance" (e.g., Wilson 2017). In many cases of physical modeling, it is a good strategy to efface or avoid detailed representation of some features of the system being modeled in order to highlight other features that receive more detailed attention. Thus, a great deal of complex physics may be summarized in a small number of macroscopic parameters, whose values are measured experimentally rather than derived, and this information then used in more detailed modeling elsewhere. Or certain gross features of complicated physics may be captured via a particular choice of boundary conditions, with these chosen so that they will fit with (allow for the solution of) a particular set of differential equations that model what is going on elsewhere in the system such as its interior, with the latter being modeled in much more detail. Again this is a strategy with a means/ends justification that deserves more attention from philosophers of science.

2.8 Representation

I noted above that pragmatists have often been skeptical of the notion of representation when understood in terms of concepts like mirroring and correspondence—what I have called literalism. Sometimes this skepticism is global, as with Price's global expressivism. I take no stand here on this global issue but do want to argue that when one focuses on theories and models in science, rejection of expectations associated mirroring, correspondence, and isomorphism has much to recommend it.

To motivate this line of thought, let me begin with some examples. The basic claim that I want to make is that a model or theory or structure can provide information about how matters are in the world (so that we can understand the model as making claims that are *true* of the world[32]) but that we need not think of this in terms of the model mirroring or being isomorphic to what is in the world. The model tells us about the world but what it tells us need not be something we can read off by interpreting the model as a kind of mirror or isomorph of nature. Often a "literal" interpretation of a model is not what the model tells us about the world.

2.8.1 Wave Function Realism

Non-relativistic quantum mechanics models multi-particle systems with many degrees of freedom by means of a state function defined on a high dimensional configuration space that evolves in accord with the Schrodinger equation. Take this representation literally and you have what has come to

[32] Recall that by true claims I'm including claims nearly true or true enough as to make no difference.

be called wave function realism: reality itself is a high dimensional configuration space inhabited by some wavelike thing, as advocated by Ney and Albert (2013) and others. This is construing quantum mechanics in accordance with what I have called literalism—everything in the theory has the job of referring in a very direct way to elements of reality.

There are technical problems with wave-function realism but even before getting to these, a natural reaction, which I hope at least some of you will share, is that taking QM seriously does not require this sort of literalist construal of what it tells us. QM certainly tells us something about what nature is like, including so-called unobservable aspects of nature, but for this to be true it is not required that everything in the Hilbert space (or some alternative) representation has a direct counterpart in nature or stands in some relation of isomorphism to the world's constituents. It isn't even required that there be a no-trivial partial isomorphism of some kind between the theory and some subset of the world's constituents.

2.8.2 Continua

Many aspects of the behavior of gases or systems consisting of gases, liquids, and solids are well modeled by treating these as continua with an infinite number of degrees of freedom. Indeed, it appears that an adequate treatment of a number of phenomena, including phase transitions, requires such a treatment (taking the continuum limit and so on). If we approach such models with literalist expectations, it is hard not to be puzzled by this. On the one hand, of course gases etc. are not literally continua but rather are made up of molecules. We might say that the continuum models are strictly speaking false are but "useful fictions" or "idealizations" or "approximations" but in my view these are just labels that give us little insight into what is going on. On my view, often a better strategy is to think of the continuum models as telling us things that are genuinely true of (or informative about) the systems modeled, but not as telling us what we might naively take them to be saying on a straightforwardly literalist construal. That is, the best way to understand these models is not as telling us that matter literally is continuous (everyone knows this to be false) but rather as functioning so as to convey other sorts of information. For example, one thing we learn from the use and success of such models is that the bulk behavior of many macroscopic systems is surprisingly independent of the details of the behavior of their molecular constituents—so independent that even a continuous fluid will exhibit the same macroscopic behavior as its atomic counterparts as long as it satisfies certain generic constraints. Another related point is that phenomena like phase transitions (as well as many other macroscopic properties) are scale-dependent in the sense that their existence is tied to certain size and energy scales and we can think of continuum limit is a way of bringing this out.

Once again, my concern is not with the details of this construal but rather with the more general strategy of interpretation: the continuum models are telling us something true about the world (providing worldly information) but what they are telling us may not be best understood in terms of the assumption that there is an isomorphism between such models and nature. Or at least the appropriate conclusion is that we can't just read off just from the surface semantic features of these models what it is that they are telling us about how matters stand in the world. They inform us about the world but in a more subtle and indirect way than is suggested by the isomorphism picture.[33]

To make a connection with earlier portions of this chapter, one of the things that this sort of literalism leaves out is the role of the subject or user of the model. It is the user of QM who knows how to use the Hilbert space formalism to extract or capture information about the behavior of quantum mechanical systems, predict their future behavior, and so on. Similarly for the user of continuum models. Such users know how to employ these models for various purposes without necessarily construing them according to literalist expectations; users know what parts of a model should be taken literally in certain applications and which should not. Pragmatic philosophers of science should accordingly focus on how such models are used and resist the temptation to automatically construe them in literalist ways. Or to put matters differently, rather than trying to understand the use of models in terms of a relationship of correspondence or isomorphism between model and world, think instead of the models as a devices or tools that subjects use to extract or capture information about nature.

[33] To be clear, I do not mean to claim that every feature of a useful model or theory can be construed in the way described—that is, as providing true information about nature although not necessarily the information that is suggested by a literalist construal. Some predictively useful models contain assumptions that likely can't be reconstrued in this way—e.g. models of the nucleus that attribute a shell structure to it (Morrison 2015). My point is rather that many of the examples of model and theory assumptions that philosophers have regarded as "fictions" or falsehoods" need not be construed as such. Ironically, "models as fictions/falsehoods" views often tacitly adopt the very literalist expectations that they officially reject.

2.9 Realism vs. Instrumentalism

Ideas of the sort just described are frequently denigrated as "instrumentalist" in contrast to realism, which is thought to be Good. However, as I see matters, these ideas do not amount to traditional instrumentalism as that doctrine is understood by philosophers of science. Instrumentalism as traditionally conceived involves commitments like the following: (i) a sharp distinction between observational and theoretical claims and the idea that theories are just devices for calculating observational consequences, with theoretical claims not being truth apt at all and merely playing the role of facilitating calculation of relations between observational claims. Along with this many versions of traditional instrumentalism are committed to (ii) the idea that theories and models merely codify or summarize relations involving "regularities" among observational claims and do not explain (they merely describe) and do not provide genuine causal or modal information, in any sense that goes beyond the description of regularities. The anti-literalist view of theories and models sketched above not does endorse either of these commitments. Theories and models are seen as embodying modal information and as explaining and much of the information they convey has to do with matters that go beyond straightforward observation. For example, on the construal suggested above, when continuum models are used to capture information about independence from lower level molecular details this information both has modal import and is not purely "observational" in the sense favored by classical instrumentalists.[34]

Thinking of matters this way, we should be able to see that the traditional dichotomy between scientific realism (which is often formulated in a way that it involves a commitment to something like a picturing or isomorphism conception of theories or models) and instrumentalism, when construed as committed to (i) and (ii), is far from exhaustive—there are other possibilities, including the one I have tried to describe. I believe that much of the appeal of strong forms of realism understood in terms of correspondence derives from the thought that the only possible alternative to it is traditional instrumentalism. However, as I see it there is no reason why a pragmatist philosopher of science should think that if she wishes to reject forms of

[34] For further defense of this view, which I have called "instrumental realism," see Woodward (2003) and for sympathetic recent discussion see Saatsi (2019). I acknowledge that this label is not ideal, but can't think of anything better. "Effective Dependency Relation Realism" does not exactly trip off the tongue.

realism that are committed to picturing and isomorphism, she needs to take on commitments like (i) and (ii). Pragmatists should not allow themselves to be browbeaten by vague accusations of instrumentalism and they should be suspicious of formulations of scientific realism that depend on strong versions of correspondence and isomorphism.

2.10 Dependency Relations, Ontology, and Model Pluralism

In thinking about particular formulations of scientific theories (and models) we can often make a rough and ready distinction between (1) what the theory tells us about dependency relations (how certain magnitudes or properties depend on others) and (2) the entities or objects that the theory seems to tell us exist (the "ontology" or ontologies associated with theory), at least on a literal interpretation.[35] Roughly speaking, dependency relations are encoded in the laws or causal generalizations of a theory. As an example of (1) Newtonian gravitational theory tells us about how the motions of bodies depend on their masses and the distances between them. As an example of (2) Newtonian gravitational theory can be understood as claiming that there exists a force, gravity, that acts instantaneously at a distance. It is an interesting fact that theories that are the same or equivalent in what they say about (1) can be very different in what they say about (2). For example, in the Newton–Cartan version of Newtonian gravitational theory, the role of the Newtonian gravitational force is re-interpreted in terms of spacetime geometry. The ontology changes dramatically but the dependency relations are the same. Another interesting fact is that at least in many cases there is far more continuity and preservation of content between successive theories at the level of (1) than at the level of (2). General Relativity, a successor theory to Newtonian gravitational theory, allows for recovery of the laws of the latter theory in the weak/slowly changing gravitational field/

[35] Cf. Maudlin (2019, xi). "A physical theory should clearly and forthrightly address two questions: what there is, and what it does." Of course in recommending that philosophers should focus just on the second question at least when it comes to understanding scientific progress, my view is very different from Maudlin's. I recognize that a natural thought is that Maudlin's two questions are not very readily separable but I don't think this is right, at least in many cases. I might, for example, use non- relativistic quantum mechanics and the Born rule to show how the probability of a particle penetrating a potential barrier depends on the potential and on the kinetic energy of the particle, with this being non-zero when the energy is less than the potential, thus answering w-questions about the probability of penetration. Doing this counts as an explanation by my lights but it does not require answering questions about the ontology of the wave function.

low velocity limit. But the apparent ontology of GR is very different from the ontology of Newtonian gravitational theory. In turn GR is widely expected to break down at sufficiently small length scales (the Planck length) and presumably the ontology of a full quantum mechanical theory of gravity will be very different from that of GR. Nonetheless one expects that the dependency relations postulated in GR will continue to be correct in other, more "macroscopic" domains of application.

I take these observations and others made above to suggest (I don't claim that they conclusively establish) that insofar as (1) and (2) can be separated it is often claims about (1) that are the more secure and important forms of scientific knowledge, at least insofar as we are concerned with physical theories that have a relatively fundamental status.[36] It is clear how information about dependency relations contribute to predictive success. And insofar as explanation has to do with correctly answering what-if-things-had-been-different questions (w-questions), as argued in Woodward (2003), it is again (1) rather than (2) that contributes to explanatory success.[37]

The gravitational examples immediately above involve fundamental physical theories but I also think that at least one important feature of the examples fits a great deal of science outside of physics. This feature has to do with the domain or scale dependence of successful theorizing in many areas of science. That is, in many cases we have collections of different theories or models of what in some sense are the same systems, each with their own

[36] I don't mean that the ontology associated with a theory is unimportant in science. It may, for example, help to facilitate visualization and calculation as with various mechanical models of the ether. It may also help with extending the theory to new applications. But this does not require taking the ontology as a literal truth that will be preserved across theory change.

[37] Thus on this picture, getting the ontology right (whatever that might mean) counts for less in successful explanation than getting the dependency relations right. Note also that on this view, "truth" or something in the neighborhood of truth matters for successful explanation. The question is: truth about what? I suggest it is truth about the dependency relations holding in a particular domain, rather than truths about ontology.

It is also worth noting in this connection that many prominent arguments (e.g. Laudan) for the "pessimistic induction" and similar conclusions focus on failures of preservation of the ontologies of successive attempts at fundamental theories: it is noted that particle theories of light are replaced by theories according to which light is a longitudinal wave, then by theories that take light to be a transverse wave, each accompanied by changing accounts of the mechanical ethers in which these "vibrations" occur. These ontology-related features of scientific change to which Laudan and others point are quite real (as noted above) but this is consistent with the observation that there is accumulation of increasing knowledge of dependency in optics, as noted by Saatsi (2019). It is ironic that defenses of the existence of scientific progress and the cumulative character of scientific knowledge have very often been in framed in claims about terms of continuity of reference for "unobservable entities"—that is, as claims about preservation of ontology. If my argument above is correct, this is not where one should look for continuity and progress in science.

domains or regimes or proprietary explananda. These theories/models work well (in terms of getting the dependency relations right or almost right) in their proper domains of application but either break down in the sense of yielding false predictions outside of their domains or else are unable to make predictions at all for reasons having to do with computational and other sorts of epistemic limitations. They are thus "effective" theories rather than theories of everything. Often not just in physics but elsewhere, the domains in question are distinguished by means of scale-based considerations—either spatial, temporal or energetic. For example, Herz et al. (2006) describes a number of models, at different (spatial) scales or levels, of neuronal behavior. Fine-grained multi-compartment models divide the neuron into a number of different "compartments" and attempt to understand how the details of its spatial structure contribute to certain features of its behavior. Computational and analytic impossibilities make it impossible to use such models to predict or explain qualitative features of the behavior of whole neurons such as the generation of the action potential. Instead a very different single compartment model such as the Hodgkin–Huxley model may be used for this purpose. Herz et al. also describe models at additional levels (a total of five) both "above" and "below" that of the Hodgkin–Huxley model.[38]

Pragmatist philosophers of science need not regard this (multiple models at different scales, each with their own proprietary explananda) as a troubling state of affairs. A theory or model can do a very good job of capturing dependency relations (and explaining, at least on a w-conception of explanation) within a given domain or a given scale, even if it is not successful at others. Theories/models of the same system but at different scales or designed to capture different features of those systems thus can viewed as complimentary to the extent that (as is typically the case) they make predictions about/provide explanations of different explananda, rather than conflicting predictions about the same explananda—the models of neuronal behavior at different scales mentioned above provide a good illustration of this.[39] Of course, as noted above, it is sometimes (but by no means always true) that to develop an empirically adequate theory of a given set of

[38] For more details, see Woodward (2018).

[39] In other words, as I understand "conflict" it arises when two different models make inconsistent predictions about what will happen under the same conditions—model 1 says that under condition X, Y will result and model 2 says that under condition X, Z≠Y will result. The point that models at different scales are often complimentary is developed in detail in Morrison (2015).

explananda at a particular scale one needs *some* information from other scales, so that getting theories originally designed for different explananda to talk to one another becomes an important problem. ("Some information" does not mean that all of the information provided by a model at another scale is needed or relevant—this is rarely the case.) Recognizing this is different from saying that a plurality of models is in itself objectionable or that we should always attempt to replace these with a single theory of everything.

So far I have focused on pluralism of theories/models in connection with dependency relations, where I suggest that matters often go relatively smoothly, in the sense that there is complementarity rather than conflict. The problematic issues seem to be the ones that arise in connection with (2)—the ontological commitments of the models, at least in important parts of physics. Here it is tempting to see conflict or outright contradiction rather than complementarity. If, say, we use the Navier–Stokes equations to model aspects of fluid flow, aren't we assuming that the fluid is a continuous substance of some kind? How then can we consistently also model the fluid as composed of molecules? How can we model spacetime as flat with a gravitational force that acts at a distance but also curved with no such force?

One way of responding to such conflicts is to argue that in each case just one of the conflicting ontologies must be correct—typically this will be the ontology associated with the more fundamental theory. Thus one might conclude that continuum models of fluids and gases are "false" because they involve commitment to a false ontology and so on. This in turn generates the various puzzles, gestured at above, about how false models nonetheless seem to be illuminating (or perhaps leads to a denial that they are illuminating). Rather than wading yet again into these waters let me just reiterate the observation that the ontologies that go with many theories seem to be the least stable and secure part of those theories and arguably the parts that do the least work when it comes to prediction and explanation. Perhaps, then, there is something to be said for taking such ontological commitments less seriously or having a more relaxed and permissive attitude toward them. To the extent that one does this, conflicts in ontology will seem less alarming.[40]

[40] To be clear: of course I don't deny that there are "ontological" claims made by scientific theories that are secure and unproblematic: we are not going to change our minds about the existence of atoms and molecules or the chemical composition of DNA. The philosophical debates about "ontological status" typically arise for other sorts of claims, often but exclusively for theories that are taken to be "fundamental," as in the examples discussed above. It is these cases where we don't seem to find continuity at the level of ontology. Above the fundamental level, we seem to have "effective" or "scale-dependent" ontologies, that are often preserved across theory change.

2.11 Modality Again

One of themes illustrated by the examples above is that we should not identify the issue of whether a claim or model conveys correct information about how matters stand in the world and whether it can be assessed accordingly (that is, as true or false, accurate or inaccurate, and so on) with the issue of whether that claim or model stands in a mirroring or correspondence relation to nature.[41] The assumption of correspondence leads one to look immediately for things or entities or structures that are the worldly relata of the correspondence relation—high dimensional Hilbert spaces as "real" and so on. The pragmatist thinks instead that the mathematics of Hilbert space can be used to say true things about quantum mechanical systems without its functioning in such a simple mirroring sort of way.

Let me conclude by floating the possibility that a similar point might hold for the use of modal concepts in science. (I readily acknowledge that what follows is more like what Richard Feynman once called an idea for an idea, rather than anything worked out.) In thinking about the physical modalities, I have always been struck by two considerations that seem prima-facie to pull (or at least pull me) in opposite directions. On the one hand, it is very hard to deny that claims about physical possibility (and perhaps even more so, claims about physical and causal dependence) are the sorts of claims that are true or false and that they are in some sense answerable to the world. For example, in my view it is very hard to understand what could possibly be going on when one infers to causal conclusions on the basis of experiments if this were not the case and if the modality associated with causal claims was simply a "projection" of our inferential practices onto the world. Moreover, attempts to reduce modal claims to non-modal claims about regularities and so on seem uniformly unsuccessful. This may seem to point us in an anti-Humean direction. However, on the other hand, attempts to accommodate the observations just described through the postulation of "non-Humean stuff" (John Earman's terminology)—powers, dispositions, relations of necessitation among universals, and so on) seem unilluminating. As suggested above, they don't provide ordinary scientific explanations of anything and don't provide guidance for how we should test or reason about causal claims. Moreover, at least in many cases, this

[41] For a treatment of modality and causation in a somewhat similar spirit, that emphasizes the role of phase space in representing possibilities, see Andersen (2017).

postulation leads to all sorts of puzzles of a placement sort—about the "ontological status" of these non-Humean things and relations.

My suggestion is that perhaps there is a way to accommodate both of these reactions by analogy with the treatment of models and theories floated previous sections. I assume that there is nothing wrong with the idea that modal and causal claims are truth apt and that, when true, are informative about the world. Where we go wrong is when we try to understand the truth-aptness of modal claims in a literalist or object naturalist fashion—that is, by postulating special things or entities (or entity-like properties) to serve as correspondents for such claims. Thus while it indeed may be true that (i) it is possible that I might be now standing two feet to the left of where I am now standing, we should not try to interpret this claim in terms of the existence of some special entity such as a possible world in which I am two feet to the left. This misunderstands the way that (i) works, construing it on the model of something like "the cat is on the mat" (where the referents of "cat" and "mat" are unproblematic and even the "on" relation has a straightforward correspondent) and then inventing special new stuff to accommodate this construal.

Similarly, (ii) "aspirin causes headache relief" is unproblematically true, but I think that it is consistent with this to doubt that "causes" has the function of naming some special sort of entity—the "causal tie"—which we might go on to isolate and investigate in the way that we might investigate, say, cats, electrons or the Higgs boson. Talking about causation in the context of a claim like (ii) is better viewed as (something like) a claim about what would happen if one were to ingest aspirin in certain circumstances (in cases in which other causes of headache relief are not present, etc.). Of course there are conditions in the world at a more fundamental level (blocking of prostaglandin production and so on) that support this and other causal claims but these conditions are often complex and spatially and temporally distributed in ways that may not be immediately transparent. Moreover the conditions in question may be non-trivially different for different causal claims. This is why, although one finds plenty of discussion of causation, methodologies for causation inference and so on in science, one does not find investigations into the kind of stuff or thing causation is in the sense one finds such investigations for, say, electrons or DNA. Put differently, "causation" is a fit subject for methodology but not necessarily for ontology, at least if one has the usual expectations of what an ontology of causation would look like.

Let me conclude with another analogy which brings together some of themes about literalism as applied both to theories and modal claims.

Consider a 6n phase space representation of an n-particle system conforming to the laws of classical mechanics, with the representation in terms of the three position and momentum variables for each particle. Those with an inclination toward literalism may interpret this representation (or at least the claim that it is used correctly to represent some physical system) as the claim that there "exists" an abstract space having the features of the phase space—it has 6n dimensions and so. (In using this representation we are "ontologically committed" to such a space.) One might then worry about how this space relates to the more familiar three dimensional space in which we apparently find ourselves or to other abstract spaces that might be constructed. The attitude I am recommending allows us to use the phase space representation and to think of it as conveying information about what the represented system is like (in particular its modal structure) without interpreting it in this way. The use of this representation will be appropriate if for example the system represented really does have 6n degrees of freedom, if the position and momentum variables for each of the particles can vary independently of each other and so on. It is claims like this that are implied by the use of the phase space representation, not the claim that space "really" has 6n dimensions. Note that in this way the phase space representation encodes information about physical modality and possibilities but it does not do so by postulating possibilia or the like.

2.12 Conclusion

I've tried to sketch some elements of a possible pragmatist philosophy of science. However, my overall goal is not so much to convince readers on particular points but rather to point to a path forward: there are many issues and problems that belong to general philosophy of science that can be approached in a way that does not involve ambitious metaphysics. I'd be very pleased if readers were convinced of that general point even if they disagree with the details of much of what I have said.

Acknowledgements

This was written as a sort of position paper for the PragMAPS II conference at the University of Oslo in 2016. It was my attempt to engage with a number of the themes that were to be discussed at that conference. The idea was to get some positions on the table which could then be discussed. This helps to account for the somewhat freewheeling character of the paper (it was intended to provoke), as well as why I chose to refer to some ideas (discussed by other participants at this and other PragMAPS conferences) and not others that are equally worthy of discussion. In addition to PragMAPS participants, I would like to thank Bob Batterman, Porter Williams, and Mark Wilson for comments on an earlier draft and Holly Andersen for very detailed and helpful comments on the present version.

References

Akaike, H. (1974). A new look at the statistical model identification. *IEEE Transactions on Automatic Control*, *19*(6), 716–23.

Andersen, H. K. (2017). Patterns, information, and causation. *The Journal of Philosophy*, *114*(11), 592–622.

Armstrong, D. (1983). *What is a law of nature?* Cambridge University Press.

Batterman, R. (2013). The tyranny of scales. In R. W. Batterman, ed., *The Oxford handbook of philosophy of physics*. Oxford University Press, 255–86.

Bhogal, H. (2020). Humeanism about laws of nature. *Philosophy Compass*, *15*(8), e12696.

Carroll, C., & Cheng, P. (2010). The induction of hidden causes: Causal mediation and violations of independent causal influence. In *Proceedings of the Annual Meeting of the Cognitive Science Society* (Vol. 32, No. 32).

Danks, D. (2015). Goal-dependence in (scientific) ontology. *Synthese*, *192*, 3601–16.

Dupre, J. (2012). *Processes of life: Essays in the philosophy of biology*. Oxford University Press.

Glymour, C. (2001). *The mind's arrows*. MIT Press.

Godfrey-Smith, P. (2015). Pragmatism: Philosophical aspects. In J. Wright, ed., *International Encyclopedia of the Social and Behavioral Sciences*, 2nd edition (2015), Vol. 18. Elsevier, 803–7.

Goldenfeld, N., & Kadanoff, L. P. (1999). Simple lessons from complexity. *Science*, *284*(5411), 87–9.

Herz, A., Gollisch, T., Machens, C., & Jaeger, D. (2006). Modeling single-neuron dynamics and computation: A balance of detail and abstraction. Science, *314*(5796), 80–5.

Hitchcock, C. (2012). Events and their times: A case study in means/ends metaphysics. *Philosophical Studies*, *160*(1): 79–96.

Janzing, D., Mooij, J. Zhang, K., Lemeire, J., Zscheischler, J., Povilas, D., Steudel, B., & Scholkopf, B. (2012). Information-geometric approach to inferring causal directions. *Artificial Intelligence*, 1–31.

Kelly, K. (2007). A new solution to the puzzle of simplicity. *Philosophy of Science*, *74*(5), 561–73.

Lewis, D. (1986) *Philosophical papers*, Vol. 2. Oxford University Press.

Loewer, B. (2012). Two accounts of laws and time. *Philosophical Studies*, *160*(1), 115–37.

Maudlin, T. (2019). *Philosophy of physics: Quantum theory.* Princeton University Press.

Mitchell, S. (2002). Integrative pluralism. *Biology and Philosophy*, *17*(1), 55–70.

Morgan, S., & Winship, C. (2015). *Counterfactuals and causal inference* (2nd edition). Cambridge University Press.

Morrison, M. (2015). *Reconstructing reality.* Oxford University Press.

Ney, A., & Albert, D. Z. (Eds.). (2013). *The wave function: Essays on the metaphysics of quantum mechanics*. Oxford University Press.

Pearl, J. (2000) *Causality: models, reasoning and inference.* Cambridge University Press.

Pearl, J. (2018). *The book of why*. Basic Books

Price, H. (2011). *Naturalism without mirrors.* Oxford University Press.

Ruetsche, L. (Forthcoming) Pragmatism, perennialism, and the physics of ignorance [this volume.]

Saatsi, J. (2019). Realism and explanatory perspectives. In M. Massimi and C. McCoy, eds., *Understanding perspectivism: Scientific challenges and methodological prospects.* Routledge.

Schulte, O. (2017). Formal learning theory. *Stanford Encyclopedia of Philosophy.*

Schulz, L., Kushnir, T., & Gopnik, A. (2007). Learning from doing: Intervention and causal inference in children. In A. Gopnik and L. Schulz, *Causal learning: Psychology, philosophy and computation.* Oxford University Press.

Sider, T. (2011). *Writing the book of the world.* Oxford University Press.

Spirtes, P., Glymour, C., & Scheines, R. (2000). *Causation, prediction and search.* MIT Press.

Strevens, M. (2008). *Depth: An account of scientific explanation.* Harvard University Press.

Wilson, M. (2017). *Physics avoidance.* Oxford University Press.

Woodward, J. (2003). Experimentation, causal inference, and instrumental realism. In H. Radder, ed., *The philosophy of scientific experimentation*. University of Pittsburgh Press, 87–118.

Woodward, J. (2013). II—Mechanistic explanation: Its scope and limits. In *Aristotelian Society Supplementary Volume* (Vol. 87, No. 1, pp. 39–65). Blackwell Publishing Ltd.

Woodward, J. (2014). A functional account of causation; or, a defense of the legitimacy of causal thinking by reference to the only standard that matters—usefulness (as opposed to metaphysics or agreement with intuitive judgment). *Philosophy of Science, 81*(5), 691–713.

Woodward, J. (2016). The problem of variable choice. *Synthese, 193*(4), 1047–72.

Woodward, J. (2021a). *Causation with a human face: Normative theory and descriptive psychology*. Oxford University Press.

Woodward, J. (2021b). Explanatory autonomy: the role of proportionality, stability, and conditional irrelevance. *Synthese, 198*(1), 237–65.

3
Trueing

H. K. Andersen

3.1 Introduction

What is it, exactly, that makes it worth declaring oneself a pragmatist, as a practicing philosopher of science? What does pragmatism offer that could be worthwhile if one is (like many philosophers of science, and some pragmatists as well) averse to declaring oneself using -isms? The goal of this chapter is to clarify one way of being a pragmatist about truth even if one does not work on anything particularly related to truth. Our background views on what truth is like still shape our other views about, for instance, idealizations, explanation, and understanding. Pragmatism offers a way of un-entrenching some entrenched (and, thus, less fruitful and more frustrating) debates, and re-orienting them in ways that improve the possibility of making genuine progress. This means my arguments here are more synoptic and sweeping than thorough and precise, with the aim of convincing that there is something to this pragmatist approach rather than spelling out every detail.

Part of my aim is to highlight the inextricability of questions that count as part of metaphysics and questions that count as part of epistemology. While this volume is directed toward pragmatism at the intersection of metaphysics and philosophy of science, this chapter supplements this with some pragmatist epistemology as it fits into these pragmatist metaphysical views. Addressing metaphysical issues without considering the deeply connected epistemological issues will always result in a partial or distorted view. Similarly, questions about epistemology cannot be neatly hived off from metaphysical presuppositions and implications. Substantive answers to metaphysical and epistemological questions need to occur in an integrated and mutually supportive way to be illuminating of knowledge practices and products in the sciences.

H. K. Andersen, *Trueing* In: *The Pragmatist Challenge: Pragmatist Metaphysics for Philosophy of Science.* Edited by: H. K. Andersen and Sandra D. Mitchell, Oxford University Press. © H. K. Andersen and Sandra D. Mitchell 2023. DOI: 10.1093/oso/9780198805458.003.0003

Even when one is working in areas of philosophy of science that don't involve formal treatments, one's background view of what truth is, and how it behaves, still centrally shapes one's other views on issues. Everyone, not just those doing formal work, needs to have at least a moderately clear and appropriate understanding of truth in order to effectively do a wide range of work. There are pragmatist ways of thinking about truth that can be much more useful in thinking about how models and larger theoretical units of commitments, including instrumentation, fit onto the world. We need not endorse a sharp distinction between truth and justification, or an atomistic approach to assigning truth values or credences to scientific results, or a distortion of holistically connected concepts and claims, to nevertheless have our views shaped by the background implicit view we hold about the nature of truth. Epistemological norms continue to play foundational roles even in views that purport to reject truth as a central aim of scientific work. I claim that we are better off being explicit about how we think about truth, and improving that view, rather than rejecting it and reinventing the wheel from scratch with epistemic norms under different names. James and other early pragmatists rejected a correspondence theory of truth. In contemporary discussions, many views offering understanding as some replacement for truth or for explanation also reveal that what is being rejected is a *particular* view of truth, not truth *tout court*. We could instead reject that implicit but guiding view of truth, which is a correspondence theory applied atomistically, in favor of a more apt way of thinking about epistemic norms in the sciences.

Thinking of truth as trueing, like trueing[1] a bicycle wheel, facilitates thinking of models and products like diagrams or maps in terms of how well-trued they are to their target. It captures the Duhemian holism that is also central to pragmatism. It emphasizes the *process* by which models are brought into true with their target systems, and the fact that this process is not accomplished once and then is done forever, but instead requires upkeep and ongoing fine-tuning even in the best of cases. True need not be thought of as part of a binary state ascribed to propositions corresponding to individual bits of the world, true forever or never. True should instead be thought of in terms of being *in true*. It is a directed relation between a knowledge product and some feature(s) of the world. It is not ascribed only to individual propositions, or even to well-defined collections of

[1] Ordinarily, this term should be spelled "truing." However, I am retaining the "e" to emphasize the connection with truth in this philosophical re-use of the original term.

propositions. It admits of degrees and can be applied to a wide variety of formats, including maps, diagrams, highly idealized models, and more. There is no single Right Way to true a wheel, though there are clearly norms by which we can judge if a wheel is in true. This is not so much of a new, water-tight pragmatist account of truth. It is a Jamesian metaphor applicable in scientific contexts where formal approaches are not at the forefront, to make truth-apt norms clearer and more format-appropriate for application in different sciences.

A characteristically pragmatist emphasis is on the *connection* between belief and action: beliefs are for guiding actions, and actions are guided by beliefs (inter alia, Misak 2013). These two must fit, and knowledge processes and products (beliefs, methods of justification, factual knowledge derived through application of these processes, etc.) are normatively evaluated by this fit. Furthermore, the beliefs in question are not merely individualistic. The beliefs and actions in question are themselves social—no single individual's beliefs are at issue, but instead the shared belief-like structures of effective working models and other knowledge products. Even when philosophers of science reject "truth" in their accounts of various aspects of scientific work, such as instrumentation or explanation, there is ongoing use of characteristically epistemological norms in a social enterprise where scientists communicate and coordinate with one another. Knowledge in the sciences is intrinsically collective: we inherit much of what we know without having contributed to it, nor having even checked the evidence; we generate and refine it using socially structured methods that thereby improve our resulting knowledge. This means also that a pragmatist approach to questions in philosophy of science recognizes the hopelessness of a strict segregation between various kinds of values and other aspects of inquiry. Our actions, past and future, *unavoidably* guide inquiry, not just for ourselves but for communities of knowers and actors. Committing to pragmatism can then be seen as a commitment to accept this explicitly and do a better job in providing that guidance.

In the last section, I make the stakes of accepting a pragmatist way of thinking about truth clearer. Many of us are concerned with a characteristic anti-scientific attitude, one which situates the activities of non-specialists reading misleading pop articles as "research," one which rejects scientific results with challenges of the form, "But do we really *know* that?" This challenge is one we lose when we cede ground by acknowledging that we do not *know*, for instance, that human-caused climate change is driving increasing ecological disasters and climate instability. The current strategy is often to say, "No, we don't *know* that, but..." and then add caveats such as: science can't really know anything; here is our extensive justification for this claim; here is how knowledge generation in science works; etc. I think these are all wrong, from the start. We *do* know these things, full stop, on a pragmatist account of truth and knowledge. Hedging because we have concerns about the correspondence theory of truth, or about a speciously precise Bayesianism, means that we end up inadvertently misleading audiences. We don't have to choose between our philosophical scruples and our social ones. As a pragmatist about truth as trueing, and knowledge in the sciences involving socially held commitments rather than merely individualized personal beliefs, we can simply state, of many such questions, that yes, indeed, we *do know* this. Importantly as well, it leaves open space to say of some scientific work that no, at this stage, we do not know this, and have it be a more constructive act of communication precisely because it contrasts with other claims that we do already know. Pragmatism aptly navigates the space between a knee-jerk scientism and anti-science, anti-expertise trends. This is not merely something we *can* do with a reclamation of truth; it is something we *ought* to do, for reasons that are scientifically, metaphysically, epistemically, and socially normative.

Briefly: in Section 3.2, I describe the "no answers at the back of the book" problem for many contemporary discussions in epistemology that preclude usefulness for philosophy of science. Section 3.3 offers a short overview of highlights from William James on truth and clarification of the pragmatist challenge, "what difference does it make to experience?" Section 3.4 describes trueing as a holist, process-oriented way of thinking about truth for knowledge products like models, etc. Section 3.5 shows how several recent rejections of truth as a central aim in science are actually rejections of a particular view of truth, and instead of rejecting truth tout court, we could reject that correspondence view and replace it with a pragmatist view of truth. Section 3.6 situates trueing in terms of means–end metaphysics and epistemology, and Chang's work on the inextricability of conceptual change and instrumentation development. Section 3.7 makes the case that philosophers of science have an obligation to be pragmatists about truth.

3.2 No Answers at the Back of the Book

This chapter is not a rejection of traditional epistemology, nor a positioning of philosophy of science as somehow superior at the task of assessing knowledge. However, there are some real reasons about why traditional epistemology has left philosophy of science cold. Some issues to which traditional epistemology has paid attention rarely apply in philosophy of science, and some of the key simplifications made in service of discussions in epistemology seem like trading in the very things philosophers of science are interested in. The key point I will focus on is the emphasis on variations of the justified true belief (JTB) structure. Even when questioning or denying that justified true belief is knowledge (inter alia, Gettier 1963; Zagzebski 1994; see Ichikawa and Steup 2018 for more), the focus is still on these three elements as distinguishable. What accounts involving a JTB structure have in common is a separation between the J and T such that one could have all the justification in hand for a given theoretical claim, from all different available evidential avenues, and still be able to coherently ask, "It's well-justified, but is it *true*?" That gap between truth and justification is where one imagines that some extremely well-justified theory might in fact be false. That question doesn't simply miss, but actively undermines, the processes and dissemination of scientific work.

It may seem obvious that there is a different between truth and justification. My point is not to deny that there are many excellent examples that illustrate this difference, but instead to offer an existence claim: there also exist cases where there does *not* exist a clean distinction between truth and justification, and these are left out of or distorted by any account that insists on there always being such a distinction. In a scientific context, requiring the T and the J to be always distinguishable reveals a naive expectation I will call the "answers at the back of the book." When learning physics, as an example, one learns from a textbook in which various problems are given at the end of a chapter, and the student works these problems out on their own using the material in the chapter. It can be quite time-intensive to solve even one or two of these problems, and more immediate feedback that one is on the right track is useful before doing a whole set of problems the wrong way. A solution to this is to have a section at the back of the textbook in which the answers to selected problems are printed. Students can simply check and see if they got the correct answer. When one finishes a particularly elaborate calculation and finds that one has reached exactly the right answer, there is a distinctive elation to that success.

Textbooks don't lay out the steps in those answers back there, though. The whole point is to figure out how to solve it yourself. Because of the role it serves, the answer at the back of the book is often just a single number. If you did every step of the solution correctly, this should be the number you got. Yet it is completely context-less: finding out that the number is different than the one you got doesn't give you much to go by, in terms of what you did wrong. Any mistake, at any stage, is enough to end up with the wrong answer. It can be laborious and frustrating to work through where one might have gone wrong, and how wrong one might have gone. But in solving the problems, there is always the faith that there is *one* right answer, and there is *some* means of solving the given problem with the existing resources, specifically the resources in that very chapter, to yield the right answer. Leaving out the steps heightens the clarity of the resulting answer: you either nailed it, or you did not. There is no approximately correct here.

The T in JTB presupposes that the world has an equivalent of the One Right Answer at the back of the book. Once we cash in all our justification tokens, there is still a further question about whether we got it right. If we could just check the answers at the back, lift the veil for a moment, there would be a determinate and unique Yes, This is True, or No, This is False. This presupposition, the expectation of there being a single right answer we either got or failed to get, is ill-fitting in the context where science is genuinely expanding the boundaries of human knowledge. In these cases, what can be called the liminal cases of research, there just is no equivalent to the answer at the back of the book. There is no chance to check our answers against the "right" one to see we are solving problems correctly. And importantly, this is more fundamental than there simply being a lack of access to the answers, determinate though the answers are. It is *not* merely that such answers exist but we can't get to read them. They simply don't exist. There isn't an equivalent of the instructor's manual with all the right answers in it. There will never be a point where we get to peek behind the veil of the universe and see if we got it "right." There is no guarantee of a path to a guaranteed, unique solution via the recommended resources for the problems that face us. There is no designated set of tools to solve it that the chapter points us toward. And even as we progress in physics, these answers don't appear: in principle, they simply are not there. All the justification in the world is, bluntly, all there is: there is no further "and this is Right" that can get tacked on in these cases.

Discussions about how to analyze the J, T, and B in JTB accounts share the assumption that there is a meaningful distinction in *all* cases between truth and justification. Part of what makes the sciences distinctive from other contexts in which we say that epistemic agents know something is the liminal character of what is known. Contrast this with ordinary contexts, where we might say things like, "I know that so-and-so is at the coffeeshop."

We can question how well our evidence lends justification to the knowledge claim. We might discover that our evidence is misleading. But in these ordinary contexts, it does make sense to distinguish the truth of the matter, whether so-and-so *really* is at the coffeeshop or not, from that evidence and the justificatory support it lends the claim.

In cases to which JTB applies more straightforwardly, there are three important features. First, as noted already, there is a clean distinction between truth and justification for the proposition. Second, no other propositions, and no communities of researchers, are ineliminably involved in adjudicating the meaning of the claim involved. Third, there is determinacy to such a claim being true or false. There aren't cases where, having resolved all the possible facts and how they related to the claim at hand, we might not really be sure, after all this, if we should call it true or false. There is no community of other researchers whose consensus, or lack thereof, could impact our willingness to say it is true.

Each of these three features fails to obtain where science is adding to our collective knowledge. When researchers evaluate their current best theories, models, instrumentation, or other knowledge product, what they *do not do* is add, to all the justification and conceptual refinement, an additional claim, "and this is true." They are neither leaving it off, nor adding it. It is superfluous. If pressed to answer questions like, "This is extremely well-justified, but is it true?," there will often be genuine confusion (even annoyance) at what more the questioner is looking for. Responses to these kinds of questions often involve researchers showing how to use some part of a model, how the result was reached, explaining more about how the important equations behave, how they should be applied to some prototypical cases, etc. The knowledge product is shown to work well; it is put into action, and the results are taken to speak for themselves. This response cannot be separated into the justification bits and the truth bits. It illustrates knowing as a process and as action. It is not a list of propositions that get an extra column in which to put checkmarks for "is it true?" or its close cousin, "do we really know that?" There is no certification stamp to the effect, "we looked in the back of the book and indeed this is the correct answer."

Insofar as these liminal cases where science is developing genuinely new knowledge exist, they fall outside the purview of any account that requires a JTB structure, no matter how the J and the T are construed. There exist instances of knowledge where there is no meaningful difference between justification and truth. This is not simply a matter of pragmatic encroachment, where circumstantial factors shift the standards for what counts as sufficient justification. It is in-principle: there are cases where, no matter how one construes the threshold for evidence, there fails to be a clean distinction among the J, T, and B.

I'll briefly mention three specific examples to show how plays out in rendering contemporary epistemological discussions less helpful or worthwhile for philosophy of science: the knowledge norm of assertion; the KK Principle; and lower- versus higher-order evidence. These are not singled out because they are especially bad, nor am I claiming that there is nothing worthwhile happening in those discussions. Rather, these examples illustrate how this background assumption about JTB still infuses these discussions and makes the price of entrance too steep for philosophy of science.

According to the knowledge norm of assertion (Williamson 2000), one is only justified in asserting something one knows. In order to have genuine justification, one first has to have knowledge. Something that turns out to be untrue, then, could not have been justified at any time of assertion. Consider, though, how this would work in ascertaining whether physicists were justified in asserting the discovery of the Higgs boson. In one way of understanding justification, they were arguably more justified in this than in any other assertion in the history of science, with a fantastically high threshold of confidence of five sigmas. But one could still say, if one is a knowledge-firster, "They were only justified in asserting the Higgs boson had been discovered if they *knew* that the Higgs boson had been discovered." This is a way of saying that they were only justified if indeed the Higgs boson is a One Right Answer at the back of the metaphorical book of the world.

If it turns out that we replace the Standard Model with something else in a decade, then we must modify our assessment of this episode retroactively. Physicists will turn out not to have been justified at that time, because one can't know something that wasn't true, and thus they could not have been justified in asserting it, no matter the actual evidential situation. It will not matter how much further development was based on that discovery, or how much successful prediction or manipulation of high energy particles was involved. Yet from the perspective of the physicists making the discovery, there is no space between knowing that it had been discovered and the justification they had for asserting that it had been discovered. If that is not enough to count as knowing, then they don't know that they have discovered it and will likely never know. It becomes unknowable. If they don't know that they have found the Higgs boson, though, then what are they doing? Because they very much take themselves to be asserting with an unprecedentedly high degree of confidence that they know it. The Standard Model is a good example because a very large number of physicists expect it to be replaced, and nevertheless need to use it in the meantime. No one *knows* any particle physics, from a Knowledge-first perspective, despite its impressive track record as one of the most confirmed theories ever and its fruitfulness for discovery. Requiring that assertions in science are only justified if they are knowledge requires scientists to also know that what was reported is not ever going to change. This is an unattainable, and unreasonable, standard. No scientific results would ever count as genuine knowledge (or if they did, we would never know that we know those results; see below). As reassuring as it might be to a certain kind of epistemic existential anxiety, there will never be a guarantee that now we know something finally, and are done, and this result will never need to be modified, refined, reformulated, or replaced.

Similar circularities befall the KK principle. According to this principle, if one knows some proposition P, then one also knows that one knows it. Put into use in the Higgs boson example, P in this case could be, "The Higgs boson has been discovered." But pressing physicists on this issue would reveal that they don't meet the KK principle. Sure, they know that they found it. But do they know that they know? Physics has already given the best answer it has—what they know. Do they know more about what they know, such as the fact that they also know it? No: the totality of what they know was already included. The KK Principle asks, "Ok, you found the boson, with five sigmas of confidence. But did you *really* find it?" And any answer to this has to be something between a shrug, or a "I just said that we found it," or even a "I don't know what more you are asking for here." Posing this question to such a discovery raises the suspicions that what the inquirer is asking for is something that could not, in principle, be given with respect to any scientific result. Applying the KK Principle in science is akin to asking, Are you sure that this is it, that you will never change your mind? And a positive response to this query is not something that science can assert.

As a final example, consider recent discussions about lower-order and higher-order evidence (e.g., Kelly 2010; Christensen 2010). The idea is that some evidence bears directly on whether a proposition is true, and other evidence bears on *that* evidence but does not bear directly on the proposition in question. One might think one has good evidence for a belief, but then further evidence comes along to the effect that some of the original evidence might not have been reliable. The distinction involves a commitment to some evidence bearing directly on a proposition as true or false (lower-order evidence), and other evidence bearing instead on the lower-order evidence (higher-order evidence). On this distinction, the higher-order evidence bears on the proposition in question only filtered through the other evidence on which it bears.

The initial response from philosophers of science to the claim that there is evidence that bears on evidence might be something like, Yes, we've read Duhem. That there are further statements implicated in assessing the evidential support for some given claim is familiar. Claims in the sciences cannot be completely extricated from a web of other claims required to support it, a feature of evidential connection that is highlighted when something goes wrong. There are supplementary hypotheses that are not directly related to the one being tested, but are required to set up the experiment in the first place. Finding that something did not work indicates that one of an indefinitely long list of auxiliary hypotheses may have to be revised. None of these are higher-order, though: they are just part of the regular evidence base. There is no neat or consistent way to divide up the evidence into direct versus indirect. The degree of directness will shift depending on how thoroughly one renders inferential steps explicit.

The Duhemian initial reaction combines with the point above about having no answers at the back to the book. Holding that there is a consistent distinction to be drawn between evidence for some claim and evidence that bears on that evidence is just to forget how intertwined all evidence already is. This distinction also implicitly relies on there being some Right Answer at the back of the book, if we could just sort it out properly, such that only the evidence that goes into the Rights Steps to the solution yielding that One Right Answer should count as lower-order evidence. The best we can do in the sciences, as seen in cases like the Standard Model of Particles, may just be a really well-justified theory where, if pressed on its truth, there just is no further, clearer answer that could be given, even in principle. There is no answer at the back of book, hidden from us; it simply isn't there to do the work of ensuring that there is One Right Answer. And there is just a lot of very interconnected evidence to be weighed and sifted.

There are other excellent critiques of the various aspects of the JTB approach as it fails to help much with scientific knowledge. Dang and Bright (2021) offer compelling reasons to think that scientists need not have any of those three (see also Fleisher 2018). Brown (2015) takes up a similar challenge based on the involvement of non-epistemic values. I'm adding to those some further reasons that philosophers of science are dissatisfied with the tools coming out of contemporary epistemology in recent decades for analyzing what goes on in the sciences. In particular, accounts of knowledge that presuppose a firm distinction between truth and justification that holds even in the liminal cases will inevitably fail to capture what is epistemically distinctive about the sciences. We should not accept that these cases fail to be knowledge. Moreover, philosophers of science use epistemically normative terms frequently in contexts where explicit formal treatments of truth are not involved. Normative claims about knowledge, justification, and evidence suffuse our work. This means there is a space for a view of truth and of knowledge that serves a broader guiding role rather than merely serving as a formal apparatus into which we plug propositions and chug out truth values.

3.3 James and Truth as Making a Difference to Experience

James famously offered what arguably counts as one of the most misunderstood and quoted-out-of-context dictums in philosophy: what difference does it make to experience? This is the pragmatist challenge. A good deployment of a pragmatist challenge will be revealing; it is not an in-principle challenge that stops inquiry, but an open question which offers a path forward to resolution. This path might be a dead end: it is a useful outcome of issuing a pragmatist challenge that clarity on pseudo-questions can also be gained. But sometimes it simply clarifies what is at stake in a question or claim in a helpful, empirically substantive, way. James offered pragmatism centrally as a way of construing truth, rather than the more Peircean way of offering pragmatism as primarily about meaning (Misak and Talisse 2019). My focus here will be on truth, rather than on meaning and reference.

I want to lay out the features of experience and making a difference to experience that can be used to put together a way of thinking about truth that is not a version of a correspondence theory of truth. For reasons of space, I am developing the pragmatist alternative to correspondence views of truth, but not directly criticizing correspondence theories. There are excellent critiques of the naïve correspondence theory by expressivists such as Huw Price (e.g., 2011, 2013) on which I will draw. Price's critique is apt, and also fits neatly with the point in the previous section about there not being answers at the back of the book. In the correspondence view, as Price describes it, the meaning of words is a matter of matching a linguistic bit to the corresponding shape in the world, rather like kid sticker matching books. In these books, there are shadows or outlines, and in the back, there are pages of stickers. The matching game is to find the right sticker for the right outline and stick it in its proper place. There is only one sticker per shadow, and one shadow per sticker. There is a real tendency, in philosophy of science, to assume that truth works similarly. Much of the discussion around realism, for example, has assumed that theories are real when they correctly refer, and they correctly refer by having little sticker terms that correspond, uniquely, to some shadow-shape in the world, and it is the job of science to match the sticker to the right bit of world. Price offers a pragmatist alternative for word meaning, namely, that of expressionism, where word meaning is instead given by the full nexus of usages in which that word figures. I direct readers interested in more detail on this to Price's work and to Chang (2022).

James' challenge makes it useful for contemporary application, and his pragmatism both overlaps with and differs from more familiar views like early Carnap. If you like logical empiricism despite its shortcomings, you might love pragmatism. Misak and Talisse (2019) note that James was dissatisfied with the narrower version of the "pragmatic maxim" that Peirce offered, and "proposed a broader rendition according to which the point of pragmatism is not to dispel metaphysical nonsense, as Peirce had alleged, but rather to settle metaphysical disputes" (3). Carnap's popular earlier works are vehemently anti-metaphysical, in line with Peirce. But James' work shifts the rejection of metaphysics into something more useful and less intellectually dismissive, in a way more in line with later work by Carnap and especially with contemporary discussions of Carnap. James offers a formulation of the challenge that allows for it to clarify and illuminate, rather than merely eliminate or reject. He offers an unusual instrumentalist view of truth: truth is *for* something, to be assessed by how effectively it allows us to navigate experience. Truth is not sterile; it ties metaphysical and epistemological inquiry inextricably together.

I'll highlight several key features of this new pragmatist challenge, as clarifications and updates to James' original version. First, it involves a more expansive, holist, and active understanding of what is comprised by experience. Second, there is an expanded construal of what counts as making a difference. Third, this challenge is answerable, and can be used in practice to transform apparently in-principle or untestable metaphysical commitments into ones with genuine empirical consequences. Fourth and finally, because the body of experience which requires navigation is growing and will continue to grow, truth is processual, in need of ongoing maintenance and updating, not an Absolute that, once attained, must remain timeless unchanging forever lest it have always been false all along.

First, this challenge invokes a much broader notion of *experience* as something to which such a difference can be made. James' views of experience can be found in *Principles of Psychology* (1890), and in his later writings such as *The Meaning of Truth* (1975/1909). James took experience to be a holist process, a continuous stream without naturally occurring breaks or internal divisions. He rejected the idea that it came pre-marked into discrete content or time units. Experience is not composed out of smaller bits; the smaller bits can be identified and discussed by identifying them as part of the whole. Furthermore, experience is not simply passive; it doesn't just happen to us. James thought experience was active in several senses, including the fact that the actual content of moment to moment experience is actively directed by the volitional control that we have over attention. We *do* things, like attending to something, that direct and shape experience. Experience also involves our actions and their consequences, and how we respond flexibly to continue directing our actions toward our ends or goals. There is no neat way to separate out a passive component, in which experience "comes in" all on its own, from an active component "going out" over which we have control.

This expansive and active character of experience contrasts with the spare, atomistic view of experience from influential empiricist movements in the early twentieth century, like logical positivism or Russell and Whitehead. A Jamesian pragmatist can draw on this much richer understanding of experience, shifting the significance of saying that all meaningful claims, even those that have no directly observable component, nevertheless can make a difference to experience. A difference can be made by making a conceptual structure simpler to apply, or easier to explain, or more understandable. Applying, understanding, explaining are parts of a broadened view of experience, to which differences can be made.

Second, consider what it means to *make a difference* to that broadened understanding of experience. The pragmatist challenge originated in puzzles like that of the squirrel on a tree, encountered by James and his companions on a hike (1975/1908). The squirrel managed to keep on the opposite side of the tree as some hikers circled the tree. So, did they "go around" the squirrel, or no? James, apparently fed up with the tedious argument, tells the reader that if by "go around," one means, moved in a circle in which the squirrel is contained, then yes, they did go around the squirrel;

and if one instead means, went in a circle that allowed one to view the squirrel from all sides, then no, they did not.

What difference does a terminological dispute make to experience? A common answer is that it is *merely* terminological, and, thereby, not empirical or not substantive (this includes philosophy of science, and also metaphysics; see e.g., Sider 2013). A Jamesian approach to truth allows for differences in precise usage of language to be genuinely meaningful.[2] Even if disagreements about the precise boundaries of term usage do not have clear and immediate empirical consequences, they have diffuse empirical consequences that also matter for experience construed broadly. All conceptual usage in the sciences connects with other conceptual usage; no usage floats so free of other meanings that it can be changed arbitrarily with no impact. Such terminological disputes are often quite important in formulating new versions of a theory. They clarify how to apply concepts in practice; they are often important to improve understanding. All of these count as making a difference on the broader view of experience. This also highlights how collective the endeavor is: it is not the Quinean web of one individual changing terms to fit their idiosyncratic preferences. As Putnam emphasized, the language of science is a tool like a ship, not a hammer, the use of which requires many people in coordination.

Third, the pragmatist challenge is emphatically *not* a skeptical challenge or rhetorical move. It offers opportunities to be answered definitively. It is not the philosophical equivalent of throwing one's hands in the air and declaring, "How can you think that?!," or pressing the skeptical challenge of how it could all be an illusion or mistake. Applying the pragmatist challenge does not stymie further discussion; it opens it up for a clarification of what, exactly, the original claim, term, or idea is supposed to be. It can be answered by carefully examining the role a contentious concept might be playing in other reasoning, and how it connects to other, more settled, terms. It invokes pragmatist, or expressionist, accounts of meaning as the totality of usages, including inferential connections, in which a term is involved. The pragmatist challenge is intended to move discussion forward. To be well-posed, it must be in-principle answerable.

To illustrate this third point, consider the example of Andersen's (2018) deployment regarding Lange's (2016) account of laws versus meta-laws. Meta-laws, such as conservation laws, might explain why there are the

[2] This pragmatist stance on the significance of terminological disagreements is also in line with Chang's (2004) analyses of conceptual extension and revision.

phenomenological laws that there are, because they must be thus as necessitated by the meta-laws; or, meta-laws might be coincidences that hold because of the phenomenological laws that hold, where the phenomenological laws determine or explain why the meta-laws happen to hold. Lange claims that we should acknowledge this distinction as an open question, not settled by current evidence and open to revision pending future developments from science. Andersen (2018) uses the pragmatist challenge on this distinction: what difference could it make to experience, if a law is a meta-law constraining phenomenological laws, or holding coincidentally based on phenomenological laws? The resolution of the challenge reveals that accepting this as a meaningful distinction is incompatible with holding most versions of a Humean account of laws. Treating the world as a Humean mosaic requires commitment to the non-existence of certain kinds of laws, namely, those that constrain the ways the mosaic could be. Even insisting on the possibility that laws might constrain rather than merely describe rules out Humean mosaic views: the possibility of constraint laws just is the possibility that the Humean mosaic description is incomplete. By asking, what difference does it make to experience to allow this distinction as an open question?, we don't even need to have an answer to the specific question of whether such laws do in fact exist to see that this open-mindedness already rules out Humeanism; committing to Humeanism just is to commit to this distinction being meaningless or inaccurate, to commit to it *not* being an open question on which science might yield additional insight. This makes it an indirect yet still empirical issue. The challenge does not tell us what to do but it makes clearer the stakes of the distinction.[3]

A fourth point, which will be central in later sections, is that when James rejected metaphysics, it involved a fairly specific understanding of what metaphysics was such that it ought to be rejected. He was not rejecting any possible approach to metaphysics; he was rejecting the Hegelian idealism that was the then-current form of metaphysics. Indeed, he was much more amenable to the radically different "metaphysic" of someone like Shadworth Hodgson. James' pragmatism was an alternative to what he thought was a sterile idealism; he does not specifically invoke the term metaphysics, but pragmatism is a direct alternative to the specific metaphysical views of that time as a way of philosophizing about the same topics. James' pragmatism

[3] There are many other papers that use what I would count as a pragmatist challenge to make progress on a debate, though I am not presuming all such authors would also count their projects as of this sort or label themselves as pragmatists; for example, Jacquart (2021).

involves a deep inextricability of what we would now call metaphysical inquiry with epistemological questions. His rejection is of metaphysics shorn of its connections to other questions about knowledge and experience, shorn of its situation as part of natural philosophy. Insofar as truth is what helps us get around experience better, and metaphysics proceeds via questions like "what difference does it make to experience?," what it means for something to be true just is another way of talking about the difference that something makes to experience. The pragmatist challenge—what difference does it make to experience?—is the question of how to *use* metaphysics, done in a pragmatist vein, to make inferences, connect concepts, integrate different arenas of scientific work into a coherent single view of the world. On a pragmatist approach, neither metaphysics nor epistemology can be effectively done in isolation from one another.[4]

As part of this inextricability of metaphysics and epistemology, any answer to the question, what difference does it make to experience?, is going to be holist to some degree. The truth of claims cannot be assessed without the inferential connections between those meanings that mutually constrain them in a coherent web or network. Attempting to atomistically isolate and evaluate one claim by itself would involve having to pin the meaning of that claim down in such a way that would ineliminably involve a host of other claims and commitments.

In addition to being holist, knowledge for James is active: it requires updating or maintenance as the body of experience on which it is based and which it is supposed to help us "get around" grows over time. Truth will not be timeless and unchanging; even the vocabulary used to express claims will necessarily shift over time. Jamesian truth, conceived of as improving our ability to get around in experience, and the holistic and collective aspects of that experience, already carries with it the double commitment that something which makes no difference to experience could not be true; and that truth is *for* guiding actions, including the actions of belief formation and refinement.[5]

[4] See also Chakravartty (2017) on the inextricability of metaphysics and epistemology.

[5] There is a similarity here in the Jamesian approach when compared to, e.g. Legg's (2014) exegesis of Peirce's view of truth as a tool for the ongoing process of inquiry. As she points out, James has a more explicit focus on truth while Peirce's focus is on the process of inquiry. My claims about James in this section renders his view regarding the process more similar to that of Peirce's, yet without the focus on the end of inquiry. James leaves open the possibility of ongoing inquiry in a way that contrasts with Legg (2014).

3.4 Trueing

This section develops a pragmatist way to think about truth in the sciences for philosophers of science who don't usually think about truth. It is different than what many or most philosophers of science conceive of truth, or even think of as what an account of it could look like. I will not give a formal definition, or necessary and sufficient conditions; it will not look like a standard epistemological definition of truth. It actively precludes easy use of spurious mathematicisms. Instead, it is a background way of thinking about truth to facilitate more apt discussion on topics where truth itself does not play a central or formalized role, almost like a conceptual metaphor. Any time we invoke epistemic norms, *some* notion of truth serves as the foundation for those norms. We ought to make sure it is a good one. I'll argue it should be a pragmatist version, instead of the correspondence theory many philosophers of science implicitly rely on. We ought to make sure our background view of truth is one we considered and chose, rather than one we ended up with and haven't re-examined.

What we need is a way of thinking about truth that is itself useful. The view itself should be directly applicable. It should guide analyses of knowledge products like models across disciplines and also guide actions, including policymaking, that require knowledge derived from those products. It should emphasize the fact that knowledge products in the sciences are *used*—they serve functions and are put to use by a wide range of people, for a wide range of tasks. It should accommodate change in what counts as "true" without trivializing it: it should not be the case that anything accepted at some time counts as true at that time, but it also should not be the case that we only reach truth in science when we reach a final, never-changing and definitive "end of science." An effective view of truth should be explanatorily illuminating of the specifics of how what counts as true have changed through specific historical episodes.[6]

Briefly, we should think about truth in science not as a static property of individual propositions, but instead as a state into which knowledge products can be brought, and out of which they can fall, where truth is a process-oriented activity. It is aptly called trueing, and knowledge products like models can be in true or out of true. Trueing fills the lacuna identified in earlier sections: it does not require a clear and consistent distinction

[6] This bears some similarity to the discussion in Teller (2012).

between truth and justification, and it precludes application of truth to isolated statements taken out of context. Knowledge products like models have directions of fit, and norms governing what counts as fit give us a handle on what it means for something like a model—not individual statements or implications drawn one at a time from a model but the whole package—to be true.

An extended metaphor illuminates what trueing is and how direction of fit provides epistemic and action-guiding norms for the process of trueing and for using trued products. Consider bicycle wheels. A bike wheel that is in true spins around the wheel center with rotation only in one plane. It doesn't wobble (rotate side to side) nor hop (go up and down) while rotating. Most wheels are kept in true using a series of spokes. These spokes radiate from the center to the outer rim and are in tension: they pull the outer and center rims. Crucial to their performance, spokes are not aligned in the plane in which the wheel should rotate. They cross the plane in which the wheel needs to rotate, starting on one side at the outer rim and meeting the inner rim on the other side, alternating. It is the arrangement of spokes around the wheel and the tension in them, plus the intrinsic springiness of the materials, that keeps the wheel in shape so that it rides the way that two-wheeled bicycles should. If you consider just one spoke alone, it definitely is not aligned along the plane where the bike should rotate, and it is often aligned off that plane in an asymmetric way. In this way of understanding true, it is not opposed to false. It is instead a directed relation, a fit toward which something is guided or at which something can be aimed. In true is a state we aim at, which can be accomplished in degrees, not a binary of in-true or false. The wheel aims at that alignment, and even where it falls out of alignment, it is evaluated according to that direction of fit, of wheel onto plane of rotation.

When a wheel is out of true, it can, depending on how far out of true it is, probably still be ridden (hopefully to the shop to get it fixed). To true a wheel, it is taken off the bike (out of the context of immediate application) and mounted on a system that lets it spin so that a technician (who, for the sake of this metaphor and with some presumption, I'll call a scientist) can assess how out of true it is and where. Then, and this is also an important part of the pragmatist metaphor here, the wheel is put back into true by adjusting the spokes. It is vital to note that being in or out of true is not a dichotomous, yes-or-no situation. The same wheel can be more or less in true: barely out of true; noticeable but still rideable if you're careful; so out of true it would be unsafe to ride; so out of true that it can't even rotate within the forks of the frame. There isn't a fixed threshold such that above it, a wheel is in true, and below it, it is out of true. One wants to have a wheel that is more in true, other things being equal; and one can only get the true so good before it realistically cannot be improved on.

It is physically impossible to have a perfectly in true wheel, even in the best-case scenario. There is a mathematical limit toward which the wheel is aiming, in the sense that a technician working on the wheel has a completely unproblematic limiting case toward which the wheel is being trued. It is just not possible, with physical materials of any kind no matter how carefully engineered, to achieve a mathematically exact trueing of a wheel. This has nothing particular to say about any given wheel, but rather about how mathematical limits (the idealization that is the limit of perfectly in-true) fit onto the physical world.

It is meaningless to ask if any individual spoke is in true. The *wheel* is in true, because of the positions and tensions of the spokes. A single spoke cannot be in or out of true. It could be broken, it could be at a weird angle, it could be quite tight or quite loose. But it cannot be either in or out of true, by itself. For the wheel to be effectively in true, those spokes must, in fact, individually align in ways that are not directly following the overall orientation of the wheel in the plane of motion. It is *necessary* for such a wheel to be in true that the off-plane orientation and spoke tensions support each other so as to achieve the correct overall shape. It does not mean any given spoke (or, an idealization taken out of a model and evaluated on its own as not literally accurate) is out of true (or, "false"). There is not a unique "right" setting for any given spoke; each could be set to genuinely different tensions that would all result in a true wheel, given a different arrangement of tensions in the other spokes. One spoke in isolation is neither in true nor out of true.

Fixing a bike wheel that is out of true is a process that involves a lot of craft, experience, and handiness. It can be taught, but not simply by reading some guides. It must be learned by practitioners in a context where they actually have to engage in the task, and where they have clear conditions for success even though there are non-unique paths to achieve that success. There is no formula for trueing any given wheel, no recipe that will give the right result every time. It may involve hacks or shortcuts that work, even though no one is entirely sure why they work.[7] There isn't One Right Way to

[7] I have it on trustworthy authority that sometimes one might even just take a particularly bent wheel and whack it a few times, hard, on a counter. This is generally done out of view of

fix the spoke tension to get the wheel back into true. Wheel mechanics is a pluralist endeavor.

However, there are some very important limits on pluralism here. One can be a pluralist about the ways in which this particular bike wheel should be gotten back into true. But one can't just say that anything is as good as anything else. There really is a difference between a wheel that is in true and wheel that wobbles. The wobble is "wrong" in the sense of being out of true: the wheel should aim at being a certain shape, even if it doesn't quite manage that shape; and the further away it is from that shape, the worse it is. That toward which wheels are trued is not an arbitrary invention. There are stakes to getting it trued, and there are norms for judging whether something is in true, true enough, more or less trued compared to another wheel, and so on. Even two wheels that are both in true may differ, because one has spokes with overly high tension compared to the other.

Trueing the wheel is a requisite part of getting the whole bike to function as it should, and getting the bike to work effectively is a key part of getting where we want to go on it, safely and efficiently. Truth as trueing highlights that knowledge is not generated and then left in pristine, never-changing condition. Just as the most in-true bike wheel will need some ongoing maintenance, trueing our theories and models in the sciences involves ongoing maintenance and refinement. It is not a process of getting it right once, and then being done. Trueing is a process that must be refreshed, actively maintained. Even if no major changes or advancements have occurred for one particular model, it is still tied, in very Duhemian ways, to other models via the very meaning of the terms involved or the instruments used to measure the relevant quantities. This means that changes elsewhere can still shift a model out of true.

This makes sense in light of points from James, such as the claim that truth can change over time, and that it is relative to experience, so that what is true for one time may not be at a future time with new experience in the meanwhile. The pragmatist view of trueing provides for knowledge to change over time. A model that was in true a decade ago may have drifted out of true, and benefit from tightening up, getting better fitted onto the parts of world where it is applied. There is very good reason to reject any account of knowledge that requires us to not know anything till we reach some (probably mythical) state of final Perfect Science where nothing will

the customer, and is often quite effective, though in another sense it is a very bad idea with much to recommend against it. This also makes it an apt analogy for science.

ever change. According to pragmatism, such an End of Science would require the actual termination of accumulation of new experience. When cashed out this way, the End of Science is not an ideal toward which we could (or should[8]) even aim.

This also lets us make more nuanced claims about truth in the sciences than the scientistic "Everything science says is true" approach. Using the notion of trueing, it should be quite clear that some results, like whether we know that we found the Higgs boson, should count as true, while others, such as nutritional information regarding the health consequences of chemicals like caffeine, are still in the trueing process.

3.5 Trueing Models

In this section, I will consider contemporary discussions that are, I'll argue, already working with a more or less pragmatist view of knowledge, but doing so by apparently rejecting "truth" as a goal or epistemic norm in the sciences. Discussions of idealizations in philosophy of science almost universally proceed by presupposing the falsity of idealizations. I will take issue with this: only on a correspondence theory of truth, parceled out atomistically, are idealizations straightforwardly false. We don't need to reject truth and replace that function with scientific understanding. Doing so still smuggles in truth in implicit forms. We are better off just dealing explicitly with truth, and responding to shortcomings in how truth norms are applied in scientific contexts by improving, not eliminating, our view of truth. Instead of accepting a problematic view of truth on which central activities of science turn out to produce "falsehoods," we should reject the starting assumption that idealizations are actually false. We should remodel our view of truth, rejecting one that does not work (a naïve correspondence theory) with one that does (a pragmatist view). Then idealizations make a great deal more sense, analogous to spokes on a trued wheel.

[8] On James' way of thinking, the End of Science would be literally apocalyptic; it would just be whatever we had gotten to, before there was zero new experience accumulating, which could only happen where there simply were no more people to experience. This sounds dramatic and science-fictional, but it is a useful perspective: what we believe at the end of science is used in contemporary metaphysical discussions as a kind of orientation point, such as providing reference magnets for current terms (e.g. Sider 2013). It is a substantive presupposition that the end of science is some utopian state of knowledge; it is as or more likely to be dystopian, and not at all illuminating of what we, as historically situated humans doing our imperfect best now, ought to strive toward.

The correspondence theory of truth in this context is, roughly, that individual propositions should be carved out of models and then evaluated in terms of how well each of their individual terms correspond to specific bits of the world. Concerns about idealizations as false share this assumption, often made implicitly, that idealizations retain their meaning, without change, when isolated from their modeling contexts, and that they should be literally true descriptions of the world in order to count as true. This is a view held even by those who want to defend idealizations and their importance to scientific practice: the issue is framed as one where we must explain away the falsity, or explain their success *despite* the falsity.

This shared background assumption is what I will isolate and reject here: idealizations are not straightforwardly false. They don't need to be made more true, or accounted for as false yet useful. They are only false on a naive correspondence view of truth applied atomistically to artificially isolated sentences. These idealizations make sense, and the models in which they must be interpreted are true, on the trueing view. Situated in the context of models trued onto the relevant systems in the world to which they are applied, these idealizations are just like other statements that allow us to get around experience, construed broadly, in a more efficient way. What would make them look more "true" on the atomistic correspondence view would actually make them less true on the pragmatist view of trueing. They do not require explanation as to their ubiquity and usefulness *despite* being false, because they are simply ubiquitous and useful: they are not incompatible with or in tension with truth-seeking as an aim of science. It is equally viable to reject the view of truth that makes idealizations false and avoid having to develop a new set of epistemic norms developed just to account for them.

I'll make this case by focusing on recent work by Potochnik (2017) and Elgin (2017) as examples of the assumption that idealizations are false, and that as a result, truth should be rejected in favor of understanding as an epistemic aim of science.[9] These authors differ in the details, yet share the commitment that idealizations are straightforwardly false, that idealizations persist despite falsity because they have some other attribute that renders them more useful than they would be if they were made true, and that they

[9] These criticisms of idealizations are similar to ones made by Rice (2021), and his book offers a much more detailed and nuanced discussion of the issue of idealizations in models, their role in explanation even when those idealizations are distorting the representational target, and other ways to address this than treating them as simply false. His account differs from the one I offer here, but I take his view to be compatible with this general metaphor of trueing.

require analysis or explanation in virtue of this tension where idealizations and truth are at odds. All of this, I'll show, makes more sense when we instead think of idealizations as akin to spokes: not evaluatable as true on their own. We should instead think of idealization as analogs of orientation and tension that makes them look "false" on a correspondence theory but which, on a pragmatist approach, is how they successfully make a whole model in true.

Elgin (2017) offers a very clear example of what it looks like to accept the presupposition that truth requires an atomistic oversimplification of complex scientific claims. She holds that epistemic normativity, in which she includes truth and truth-conduciveness, is not a part of what science does, as evidenced by the extensive reliance on idealizations and other claims that she labels as obviously false (see, inter alia, p. 2). I begin with her account both because it is the most extreme example, and because she specifically identifies the target of her criticisms as a Jamesian view of truth.

> Following William James (1951), many veritists hold that our overriding cognitive objective is to believe as many truths as possible and disbelieve as many falsehoods as possible (e.g., Alston, 1988, 258; Goldman, 1999, 5; Lehrer, 1986, 6; BonJour, 1985, 9). This is the objective with respect to each and every individual contention. It is, they say, never legitimate to accept an acknowledged falsehood, even if so doing would in the long run result in the acquisition of more or a higher proportion of true beliefs than the repudiation of that falsehood would (Firth, 1981). As stated, Jamesian veritism is remarkably tolerant. It does not discriminate among true (or false) beliefs. Insofar as our goals are purely cognitive, it says, we should amass as many truths and as few falsehoods as possible. (2017, 10–11)

There are several things that go wrong here. One is that, put bluntly, James never endorsed such a view of truth, and would have spurned it completely. The view she calls Jamesian is precisely the sort of view James rejected as sterile and making no difference to experience. She claims that compounding propositions with logical operators like "or" results in a new truth, and that one can thus multiply vacuous truths ad nauseum (10–11). James explicitly denied this; this proliferation of vacuity is precisely what he wants to rule out by imposing the requirement about making a difference to experience. Elgin's target is a view of truth, however, that is common in recent decades in discussions in traditional epistemology. In that regard, I concur with some of her criticisms of that view of truth. It simply is not (emphatically) a *Jamesian* view. Authors who take this approach to truth, such as Gettier and work on the Gettier problem, reject James and pragmatism. The tradition she identifies as a "Jamesian" target is the tradition that a Jamesian would reject, and a tradition that rejects James.

Elgin offers an entire account of "felicitous falsehoods" that is predicated on the idea that "science unabashedly relies on models, idealizations, and thought experiments that are known not to be true" (2007, 1). She wants to "dethrone" truth as an epistemic norm for science. The replacement is felicitous falsehoods and "objectual understanding" (e.g., Kvanvig 2003), which is understanding of a topic or area. The key parts she takes issue with here seem to be the atomistic character of many accounts of truth (see e.g., 10, 13, 115). The move to objectual understanding is at least partly intended to correct this atomism with more holism.

What is a pragmatist response to Elgin's claims? Truth should make a difference to experience, and that alone is reason to reject those views of truth in which it does not; one does not need to also reject the entirety of truth-seeking in science. We can excise with a scalpel rather than an axe. It is more philosophically precise to reject a wrong-headed view of truth and replace it with a better way of thinking about truth, than to purport to reject epistemic normativity altogether yet continue to rely on implicit versions of it in other forms, such as objectual understanding, or the felicity of felicitous falsehoods. Unless anyone's objectual understanding counts as good as anyone's else's, then epistemic norms are playing an under-the-table role. And what her objections are about is not any possible view of truth, but just one strain; that strain can be rejected without giving up entirely on usage of truth.

Potochnik (2017) follows Elgin and argues that we don't use, or need, truth in the sciences, and offers a range of alternative features of scientific practice in its stead, in particular that of understanding. What she offers is really quite close to the pragmatist understanding of truth, though she offers it as a rejection of truth. What she presumes and then rejects is not any view of truth, but one that strongly resembles a naïve correspondence theory of truth attributed atomistically to propositions. She targets idealizations specifically, not idealized models per se: it is individual elements of models that count as idealizations, not the whole model containing such components. Idealizations are taken to be statements that are not literally accurate as descriptions of the world, and is sometimes extended to include statements that are abstract in their description, such that adding more detail about concrete particulars would thereby make the statement less idealized (Potochnik 2017, chapters 1 and 4; see especially 93). Idealizations are "assumptions made without regard for whether they are true and often with full knowledge that they are false" (42).

She frames her view in response to the tension between two claims: first, that idealizations are false; and second, that they are extremely useful and ubiquitous.

> In my view, idealizations are both rampant and unchecked in science. By *rampant*, I mean that idealizations are found throughout our best scientific products, and they stand in for even crucial causal influences. By *unchecked*, I mean that little effort is put toward eliminating or even controlling these idealizations. (41–2)

Elimination or control of idealizations would involve adding details that make for a much more complicated description. In her second chapter, she situates the rampant character of idealizations as serving a simplifying function: because we are modelling complex parts of the world, we need simplifications to render that complexity more tractable. Thus, she explains the "rampant and unchecked" reliance on idealizations as a necessary strategy for scientists in such a complex world.

Note that the very definition of idealizations here stipulates their falsity. Idealizations are not merely simplifying assumptions or elements of models: they are defined as the simplifying assumptions that are *known to be false*. On her approach, there is no neutral stance for idealizations to possibly be true, or play a truth-apt epistemic role, because they have already been defined as false. She sometimes interchangeably calls them inaccuracies (e.g., 43), and wants to explain the use of idealizations under the assumption that such use *is* the introduction of inaccuracies. Terms like "rampant and unchecked" are laden with normative judgments. As with the definition of idealizations where they are false, this is somewhat question-begging. But it is central to her framing that idealizations be false. Otherwise, their centrality in scientific practice is unsurprising, and requires no move to understanding to explain it.

Extensive reliance on idealizations in modeling is only a problem to be accounted for if those idealizations are individually false and the goal of modeling is to produce individually true propositions. Potochnik solves the tension by rejecting this second claim. I'm arguing that we should reject that first claim (though we can also, of course, reject the second): idealizations are not straightforwardly false. If they are not false, then they aren't rampant or unchecked, they are instead ubiquitous and central (see also Rice 2021). Idealizations are required to true models (similar to fitting-out models, e.g., Cartwright 1999). They are neither false nor true on their own. They should be assessed in terms of their role in bring the model into alignment, to move it in the direction of better fit.

Khalifa (2017) holds that there is necessary place for truth in explanation and understanding, contra Potochnik and Elgin, and clarifies understanding without thereby rejecting truth. Yet he also subscribes to the view that idealizations are false, and that we thereby must account for why false statements can still be so useful and widespread. His view would be even stronger if one instead thinks of truth as a pragmatist, thus eliminating the need to account for the presumed falsity of idealizations and adding explanatory resources as to why such statements are both useful and widespread. A host of other participants in discussions on models and idealizations also subscribe to this idea that idealizations are false, fictional, or otherwise untrue (e.g., Bokulich 2012). These accounts would be stronger, and often better for the authors' purposes, by removing the presumption that idealizations are individually and literally false.

Note that accounts like Bokulich's and Khalifa's are not about truth per se. They are concerned with other aspects of science, and don't need a formal definition of truth to do this work. It is instead a vague form in the background that gives the impression that idealizations are atomistically false and, thus, in need of explanation. It is for this kind of task I am offering the account of truth as trueing: it can be a different mental image that subtly shapes discussions about other topics than truth itself, in a more fruitful way than the naïve correspondence theory has. One might think this is a merely terminological dispute, especially since neither Bokulich nor Khalifa reject truth-seeking as an aim in science. Indeed, Bokulich (2016) offers something quite close in spirit to the view developed here. But terminology also matters. This is not *merely* a terminological dispute, but a difference in the ways in which we can organize, adjudicate, and find our way around within the total scope of human knowledge. That is why it is pragmatist. Better maps for how to get around the same territory are not merely terminological differences. They have consequences for action. Thinking of truth as trueing can *facilitate* these other discussions, not merely rearrange terms.

I've discussed idealizations in terms of how they help true models onto the parts of the world for which they are used. This is subtle but deliberate terminology: I am not merely talking about representation. Some models may work by successfully representing. But some work without representing; for instance, views about models as tools (e.g., Bailer-Jones 1999; Sterrett 2006) are quite amenable to pragmatism (it is likely also that Weisberg 2013, especially chapter 6, is pragmatism friendly). Models, standing in as a central example of a knowledge product, are *for* something, putting active and action-guiding aspects of scientific practice more explicitly at the center of the discussion. It is suitability for that use that settles the questions about how in-true the model is and how it could be made more so.

This complements but also refines and extends the means–end epistemology of Schulte (1999), means–end metaphysics of Hitchcock (2012), and means–ends analysis offered by Woodward (2023). Trueing helps identify the granularity at which means and ends are aptly picked out. Return to the bike wheel analogy. It is very normal for a technician to ask, for what will you be using the bike? Because the wheel needs to be in true not for some purely aesthetic Platonic ideal of perfect circles. It needs to be in true because you use it, by riding your bike to do things, and in order to do those things on your bike, the wheel has to be capable of filling certain roles effectively. But the technician is not looking for answers like: I will ride it to the store, and to the library, and also to work. Those are not well-differentiated ends for this wheel to serve as a differentiated means by which to achieve them.

The wise technician listening to those answers will instead ask a few clarificatory questions, like: Are those all paved surfaces, or are any of the paths gravel? How smooth are the roads you are riding on? Are you usually on the sidewalk or a side road, or are you also riding in heavy traffic on busier roads? How fast are you riding? Are you the main person riding the bike, or is it shared? The more specific descriptions (to work, to the store) of the end cluster together (all on the same paved sideroad), because there is no difference in the performance of the wheel that matters to your performance of them. Instead, ends like the type of surface over which you ride, and the speed at which you ride, matter a great deal to how the wheel needs to perform. Means–end analysis runs the risk of fine-graining the ends too much compared to the different means available. An overly fine grain obscures the underlying unity of means required to reach those ends. Thinking of means–ends analysis in conjunction with trueing for scientific practices and knowledge products guides the analysis to pick out ends, and thereby means, with the right granularity.

Trueing also connects to the work of Hasok Chang (see especially 2004 and 2022). Chang emphasizes how instrumentation development goes hand in hand with conceptual refinement and extension, and model or theory development. He identifies the epistemic circularities endemic to any development process in the sciences where new instrumentation is used to measure quantities. These circularities look self-defeating, on a more traditional epistemological analysis. Yet these "puzzles" don't look so puzzling when we think in terms of trueing. Chang offers an account of epistemic iteration for the dynamic of how progress can occur despite epistemic circularity. The principle of respect offers a resistive force to change, counterbalanced by the imperative of progress. Small changes in iterative steps allow researchers to start from suppositions they may already know will eventually be improved. They iteratively incrementally improve their starting position, including of the initial suppositions required to get the whole process moving. Those starting assumptions may eventually be entirely rejected in lieu of replacements, a rejection only enabled by presupposing them in the first place.

This is what the metaphor of wheel trueing picks out. One doesn't simply fix a wheel, at once, with a single go. Trueing involves changing the tension on one spoke, moving to another, and testing the effects of altering a given spoke's tension by spinning the wheel. It might be that the initial change made to a spoke eventually gets undone. Nevertheless, having started with that change was part of getting all the spokes into a tension alignment that eventuated in the wheel spinning evenly along. One need not jump to the conclusion that the original change to tension was "wrong," even if it was eventually undone; nor need we jump to the conclusion that the original presuppositions were "false," even when they are eventually modified or replaced. Instead, epistemic iteration and truth as trueing both push us toward a more dynamic and temporally evolving process of refinement and improvement. Epistemic iteration and trueing both emphasize the ineliminable diachronicity of truth-seeking processes in the sciences.

As terms shift in usage anywhere in a science, these shifts alter the meaning generally, not just in the localized area in which they occur. Change in some other areas can bring a given theory or model out of true over time. There is nothing wrong when that happens: we need not judge the original version of a model to have been lacking because it now needs to be re-trued. Instead, we can see how it was as true as it could get under the circumstances where those were the available conceptual resources. With new concepts, more precisely bounded because of subsequent developments in nearby areas, we are able to true even closer than we were before. This is James' point about the growth of the body of experience that truth is required to help us get around. This does not have to mean the wheel was out of true at that earlier time. No wheel stays true forever.[10]

3.6 What's at Stake: "But Is It Really True That..."

One might think, who cares? Why should we go back to talking about truth, even if we do a better job of it, when we can skip all that and simply talk directly about other things that matter, such as scientific understanding and explanation? In this section, I am climbing on a soapbox to tell philosophers of science that it matters a huge amount that we reclaim and make use of "truth," that we be able to clearly and definitively state that some claims are true without extraneous caveats, and that we do so without the pangs of conscience that would come if we had some non-pragmatist view of truth, or no particular view at all.

There are two primary reasons on which I will focus (there are more; but I will just address these). The first is that we don't have sole claim to the term "truth," and so cannot simply declare it to be worthless or not an aim of science, when many scientists themselves take themselves to be in the business of finding (what they want to call) truth. The second is that we do a disservice to non-scientists trying to make responsible use of scientific results when we declare that truth isn't important, or isn't what science produces, or isn't a central epistemic aim for science. Trying to substitute in something like understanding raises legitimate concerns that science is really about feelings of understanding, to be had as (or more) easily from watching videos on one's own as from struggling to grasp a complex scientific explanation. We owe people clear and effective communication, because they are largely trying to develop their own beliefs to guide their own actions in order to navigate the genuine lay of the land, not merely what they wish or feel. And we can do that kind of effective communication, without compromising our philosophical scruples, by getting rid of a naive correspondence view of truth and replacing it with a more apt view instead. We owe it to others to be pragmatists about truth.[11]

[10] This section could have responded to a much larger number of other connected and interesting projects that interact with trueing in various ways. Magnus (2012) on natural kinds is such an example, as is Massimi on perspectivism (e.g. 2012, 2018) and Ismael (2007).

[11] de Melo-Martin and Intemann (2018) offer a much more detailed and pointed discussion on inappropriate dissent in science and the ways it can fracture trust. Elliott and Resnik (2014)

On the first point: as philosophers, we have some claim to be experts on truth, and as philosophers of science, are well-situated to make informed claims about the truth-aptness of scientific practices and results. But we do not have sole proprietary claim to usage of that term. Explication projects, such as this one and the views discussed in the previous section, have some flexibility in terms of settling the precise boundaries of term usage. This is already an under-the-table pragmatist way of thinking about settling what "truth" means, with word meaning given by clarification of actual usage. Sometimes usage is ambiguous, and then explication projects work on settling it to facilitate clearer use. Terms like truth and understanding hang together inferentially; moving the boundaries of the meaning of one term implicates the meanings of other terms. Expanding "understanding" or "explanation" to take up much of the work that was previously done by "truth," even if one refrains from explicitly talking about truth, is not neutral with respect to truth.

Our explication projects must still respect how the term "truth" is used in settings outside of philosophy. Indeed, the accounts of idealization in the earlier section take it to be a specific point in advantage of their views, and a criterion for competing views, that adequate justice be done to actual scientific practice. Looking for true models, or truth about a system, is a way of speaking that is used widely and extensively outside of philosophy, and most relevantly for this chapter, in the sciences. Insofar as we take ourselves to also be accountable to provide apt and accurate construal of scientific practice, we must acknowledge that the idea of truth, and the goal of truth-seeking, is one that is widely endorsed by practicing scientists. It is definitely true that mostly, they are not seeking Truth, the absolute, never-changing, perfect and literally true version where there is a correspondence that neatly matches up bits of a description to the right bits of the world. But we can reject Truth as an apt characterization of the processes and results in science, while still having to recognize the ubiquity of truth, construed pragmatistically. One might be able to convince some group of scientists that there is no truth but Truth, and that science does not seek Truth. But presenting an alternative, pragmatist, view of truth results in a very different reception. Many scientists (though not all, it should be noted) resist the idea that what they are doing is producing a timeless, unique, literal, and perfect Truth about the world. But most of them very much endorse some version

offer an effective case for the way in which values inevitably influence science communication and policy and how transparency is a solution to this. See also Douglas (2016).

of truth with a small t, and would be unwilling to give that up as a central epistemic norm for research. Pragmatism offers truth with a small "t."

The second point in favor of endorsing a pragmatist view of truth is that it matters for genuinely effective communication. As philosophers with this legitimate claim to have expertise in "truth," we need to be able to make clear and unequivocal statements about truth-apt claims from the sciences. And, we have to do this while satisfying two further considerations: that of not adding lots of hesitant or overly convoluted caveats; and that of not having to compromise our philosophical scruples in making such statements while also feeling like we are lying to or misleading our audience. This is entirely possible if we adopt a pragmatist view of truth. It is easy to mistake this point for being merely pragmatic. But pragmatic, and pragmatist, are not interchangeable. Pragmatic is something one can be with unattainable ideals of truth on which we give up because of our compromised epistemic positions. Being pragmatist involves rejecting those unattainable ideals of Truth in the first place, and replacing them. Pragmatism isn't giving up or compromising on our goal; it is having a different goal. We can be pragmatists about truth without having to be pragmatic about truth.

It is a social obligation that we have, in communication with practicing researchers and with others who are neither scientists nor philosophers of science, that we accurately and clearly communicate without unnecessary confusion. It is just true, full stop, that vaccines do not cause autism, and COVID is not a hoax but an incredibly serious disease, that human-induced warming of the global climate is an unfolding disaster we can yet alleviate. Even though science is not perfect or complete, it is true that science remains our best method for finding out about how to address challenges like public health, like effective policies on education, like appropriate sources of information on which to base decisions and for guiding our policies and actions to best achieve our goals. The most effective way to genuinely communicate this is to be a pragmatist about truth and knowledge.

Adding caveats like "well, we can't say it's *true* that humans are exacerbating climate change, but our best theories tell us that...," or " we don't *know* that this series of heat waves is occurring because of human-induced climate change, but it is highly probable...," needlessly undermines legitimate communicative goals and conveys a misleading impression. It *mis*communicates. Even if what we wanted to convey was precise and accurate, it is not *successfully* conveyed, and we have to take that success in communication into account. It does an injustice to those views: it falsely makes them weaker, in a way that creates understandable confusion in our audience.

When we invoke more complex notions like justification or understanding as a substitute when answering questions that are about which claims are true or count as knowledge, we thereby mislead about the status of those claims. Philosophers might take themselves to be making classificatory points about preferred terminology. What is opaque to the audience about the philosophical reaction to first correct the terminology then answer the question is that someone doing this wouldn't actually endorse *anything* as true, because they reject the very idea of truth. Insofar as they would be willing to endorse any assertion as true, this assertion under question would be among those endorsed. That is why this is an inappropriate context in which to be making those terminological corrections: instead of succeeding in correcting terminology, it misrepresents the epistemic status of the claims under discussion.

Furthermore, these caveats and terminological corrections open space for pernicious pseudoscience to flourish. When philosophers give ground by saying things like, "we don't *know* X, but we have good justification for believing it," it thereby endorses a generally incorrect view of truth, one we don't actually have to buy into anyways, and then it uses that ill-fitting view of truth to draw the inference that we lack knowledge. This is backwards. We do in fact know some of these things; and we know it on a pragmatist view of knowledge and truth. We don't have to explain all of that, every time; we can just be pragmatists, and make clear statements accordingly, and then answer further questions like "what do you mean by 'true'" with details about pragmatism if such questions arise.

What I am urging here is how *essentially* misleading it is to an audience when we refrain from a full endorsement of some statements as true. Those caveats, "we don't *know* it's true..." would be appended to anything, but the audience to whom we are communicating when addressing questions like this are unaware of that and do not have the opportunity to hear us respond to those other questions to calibrate their response to our hesitancy. As such, we are misleading them by hesitating—we are inducing false beliefs by inducing or growing false degrees of doubt about these particular statements, as if they were different than other statements that could have been, but weren't, asked instead. Communication is not a solo act whereby we do our best using our preferred terminology in exactly the specialist fashion we want to, regardless of what other people think it means.

This may sound extreme, but it matters. We would do other members of our society an injustice if we answer questions like, "Is it true that this vaccine won't give my child autism?," with answers like, "Science doesn't seek truth, but we do have a good understanding of how this vaccine works..."

This leaves the legitimate impression that science is only a collection of understandings, some beliefs and some justification, but no truth or knowledge. This encourages rather than discourages the idea that, if someone watches a video or reads a meme and gets the feeling of understanding, they are doing their own "research," just like the scientists do, and who is to say what is true or false. It can harm people. We should not overlook this: people outside of philosophy do, in fact, read and respond to what we say.

Philosophers of science don't get to stipulate how words are used in non-philosophical settings, and answering questions like this is generally done in a non-academic-philosophical setting. It is very prescriptivist, and somewhat condescending, to decide that truth is a notion that no one should use, or should use differently. It is explicated, clarified, etc. in philosophy, but does not belong to philosophers alone. Hedging on whether a claim is true or not because one has philosophical misgivings about "truth" is a way of undermining the stance of anyone outside of philosophy to ask meaningful questions that we might be in a position to answer. It is a form of epistemic injustice by philosophers against non-philosophers, to decide that the idea of truth should simply be avoided and replaced with something else, and that we are only willing to answer in our own preferred terminology. Adopting a pragmatist stance toward truth lets us answer these kinds of questions, and make these kinds of assertions, in a way that is unhesitating, apt, and in alignment with our philosophical scruples.

Acknowledgments

This chapter has been in the works for years, and owes much to many great discussions and several seminars. Thanks to: Endre Begby, Eddie Cai, Hasok Chang, Mateo Ochoa Coloma, Iman Ferestade, Christopher Hitchcock, Yichen Luo, Matthew Maxwell, Milos Mihaljovic, Sandra Mitchell, Nicolas Rescher, Aaron Richardson, Sherif Salem, Jingyi Wang, and participants in the three PragMaPS workshops.

I am deeply grateful for the opportunity to live and work on the traditional unceded territories of the Coast Salish people, especially the Squamish, Tseil Watuth, Musqueam, and Kwiketlem First Nations.

References

Andersen, H. (2018). A pragmatist challenge to constraint laws. *Metascience, 27*, 19–25.

Bailer-Jones, D. M. (1999). Tracing the development of models in the philosophy of science. In *Model-based reasoning in scientific discovery* (pp. 23–40). Springer, Boston, MA.

Bokulich, A. (2012). Distinguishing explanatory from nonexplanatory fictions. *Philosophy of Science, 79*(5), 725–37. doi:10.1086/667991

Bokulich, A. (2016). Fiction as a vehicle for truth: Moving beyond the ontic conception. *The Monist, 99*(3), 260–79.

Brown, M. J. (2015). John Dewey's pragmatist alternative to the belief-acceptance dichotomy. *Studies in History and Philosophy of Science Part A, 53*, 62–70.

Cartwright, N. (1999). *The dappled world: A study of the boundaries of science.* Cambridge University Press.

Chakravartty, A. (2017). *Scientific ontology: Integrating naturalized metaphysics and voluntarist epistemology.* Oxford University Press.

Chang, H. (2004). *Inventing temperature: Measurement and scientific progress.* Oxford University Press.

Chang, H. (2022). *Realism for realistic people: A new pragmatist philosophy of science.* Cambridge University Press.

Christensen, D. (2010). Higher-order evidence 1. *Philosophy and Phenomenological Research, 81*(1), 185–215.

Dang, H., & Bright, L. K. (2021). Scientific conclusions need not be accurate, justified, or believed by their authors. *Synthese* (199), 8187–8203.

de Melo-Martín, I., & Intemann, K. (2018). *The fight against doubt: How to bridge the gap between scientists and the public.* Oxford University Press.

Douglas, H. (2016). Values in science. In P. Humphreys, ed., *The Oxford handbook of philosophy of science.* Oxford University Press.

Duhem, P. M. M. (1991/1914 first English edition). *The aim and structure of physical theory.* Princeton University Press.

Elgin, C. Z. (2017). *True enough.* MIT Press.

Elliott, K. C., & Resnik, D. B. (2014). Science, policy, and the transparency of values. *Environmental health perspectives, 122*(7), 647–50.

Fleisher, W. (2018). Rational endorsement. *Philosophical Studies, 175*(10), 2649–75.

Gettier, E. (1963). Is justified true belief knowledge?. Analysis, *23*(6), 121–3.

Hitchcock, C. (2012). Events and times: A case study in means–ends metaphysics. *Philosophical Studies, 160*(1), 79–96.

Ichikawa, J. Jenkins, & Steup, M. (2018). The analysis of knowledge. In Edward N. Zalta, ed., *The Stanford encyclopedia of philosophy* (Summer 2018 Edition). https://plato.stanford.edu/archives/sum2018/entries/knowledge-analysis/.

Ismael, J. (2007). *The situated self.* Oxford University Press.

Jacquart, M. (2021). CDM and MOND: A debate about models or theory? *Studies in History and Philosophy of Science Part A*, *89*, 226–34.

James, W. (2007/1890). *The principles of psychology* (Vols. 1 and 2). Cosimo, Inc.

James, W. (1975). *Pragmatism* (Vol. 1). Harvard University Press.

James, W. (1975/1909). *The meaning of truth* (Vol. 2). Harvard University Press.

Kelly, T. (2010). Peer disagreement and higher order evidence. In Goldman & Whitcomb, eds., *Social epistemology: Essential readings*, Oxford University Press, 183–217.

Kvanvig, J. L. (2003). *The value of knowledge and the pursuit of understanding.* Cambridge University Press.

Lange, M. (2016). *Because without cause: non-causal explanations in science and mathematics.* Oxford University Press.

Legg, C. (2014). Charles Peirce's limit concept of truth. *Philosophy Compass*, *9*(3), 204–13.

Magnus, P. (2012). *Scientific enquiry and natural kinds: From planets to mallards.* Springer.

Massimi, M. (2012). Scientific perspectivism and its foes. *Philosophica*, *84*(1), 25–52.

Massimi, M. (2018). Four kinds of perspectival truth. *Philosophy and Phenomenological Research*, *96*(2), 342–59.

Misak, C. (2013). *The American Pragmatists.* Oxford University Press.

Misak, C., and Talisse, R. (2019). Pragmatism endures. *Aeon*, November 18. https://aeon.co/essays/pragmatism-is-one-of-the-most-successful-idioms-in-philosophy

Mitchell, S. D. (1997). Pragmatic laws. *Philosophy of Science*, *64*, S468–S479.

Mitchell, S. D. (2020). Through the fractured looking glass. *Philosophy of Science*, *87*(5), 771–92.

Potochnik, A. (2017). *Idealization and the aims of science.* University of Chicago Press.

Price, H. (2011). *Naturalism without mirrors.* Oxford University Press.

Price, H. (2013). *Expressivism, pragmatism and representationalism*. Cambridge University Press.

Rice, C. (2021). *Leveraging distortions: Explanation, idealization, and universality in science*. MIT Press.

Schulte, O. (1999). Means-ends epistemology. *The British Journal for the Philosophy of Science*, *50*(1), 1–31.

Sider, T. (2013). *Writing the book of the world*. OUP Oxford.

Sterrett, S. G. (2006). Models of machines and models of phenomena. *International Studies in the Philosophy of Science*, *20*(1), 69–80. doi:10.1080/02698590600641024

Teller, P. (2012). Modeling, truth, and philosophy. *Metaphilosophy*, *43*(3), 257–74.

Weisberg, M. (2013). *Simulation and similarity: Using models to understand the world*. Oxford University Press.

Williamson, Timothy, (2000). *Knowledge and its limits*. Oxford University Press.

Woodward, J. (2023). Sketch of some themes for a pragmatist philosophy of science (this volume)

Zagzebski, L. (1994). The inescapability of Gettier problems. *The Philosophical Quarterly (1950–)*, *44*(174), 65–73.

4

The Bearable Thinness of Being

A Pragmatist Metaphysics of Affordances

Sandra D. Mitchell

4.1 Introduction

Taking a pragmatist stance toward the practices and products of science shapes our answers to central philosophical questions.[1] In this chapter, I will explicate how scientists' conceptual and representational practices work in concert with their observational and experimental ones to stabilize acceptance of scientific realism.

A pragmatist approach identifies the roles that agency and judgment play in the human reasoning and experimental practices that support claims about nature. Uncontroversially, particular goals, lofty or mundane, direct scientists to explore certain problems rather than others, focusing both the mind and the eye on specific regions for scientific investigation. As a consequence, the scope of what is deemed real is shaped by our aims and questions. I will argue that within a perspective consisting of goals, actions, and questions, what we say there *is* and what we say it *does*, is justified by the ongoing interactions among representative models, causal experience and experiment, and conceptual frameworks in reaching a fallible convergence to what is real (see also Hacking 1983; Chang 2022). I will defend an interactionist, pragmatist account to replace fundamentalist representationalist approaches to what constitutes realism of scientific theories and models. Fundamentalism about realism has been attached to two strategies, a

[1] See Andersen and Mitchell 2023. In this chapter, I treat "pragmatic" and "pragmatist" as distinct. While "pragmatic" references means–ends reasoning, or judgments in light of a goal, "pragmatist" reflects impacts on actions or practices resulting from a wide range of influences on scientific inquiry including differences in goals, features of experiential design and implementations, views on the nature of causal relationships in different contexts, etc.

Sandra D. Mitchell, *The Bearable Thinness of Being: A Pragmatist Metaphysics of Affordances* In: *The Pragmatist Challenge: Pragmatist Metaphysics for Philosophy of Science.* Edited by: H. K. Andersen and Sandra D. Mitchell, Oxford University Press.
DOI: 10.1093/oso/9780198805458.003.0004

top–down view that what is real is read off the *structures* described by the best scientific models and a bottom-up view that what is real are the *entities* presumed by experimental practice (Chakravartty 2021). I offer a non-dichotomous alternative. What we are justified in claiming as real phenomena are the affordances[2] constructed from the integration of top-down and bottom-up strategies.[3] I will illustrate how this pragmatist realism works, by appeal to a case of a complex, robust phenomenon.

4.2 Representational Theories and Experimental Data

Much progress has been made since the early logical empiricists' simple logical rendering of the relationship between scientific theories and experimental observations in terms of the conditional relationships between H-statements and O-statements. A series of elaborations of what goes into the inferential and causal interactions in experience and experiment exposed other assumptions and finer structure to describe the warrant for accepting the claims of scientific theories and models. Explicating the additional assumptions necessary for theory testing is part of the legacy of Duhem's underdetermination arguments found in *The Aim and Structure of Physical Theory*, originally published in 1914 (1991) that focused on how to connect the practical activities of an experiment with the typically mathematical descriptions comprising scientific, specifically physical, theory. Judgments, more than just logic and raw perception, are required to relate theoretical predictions to the upshot of experimental interventions. "Judgment," I argue, involves appeals to interdependent semantic, epistemological, and metaphysical assumptions. The language describing

[2] "Affordance" is a term introduced by J. J. Gibson (1979 reissued Psychology Press 2014), an ecological psychologist, in developing his non-representationalist account of perception. For Gibson, an affordance invokes what things external to an agent provide to the agent for action. For example, relative to the abilities of a given animal, a given surface may be "walk-on-able and run-over-able" (Gibson 2014, 127). Affordances are the product of both the agent and the environment in which the agent acts.

[3] While my appeal to Gibsonian affordances is in support of a version of actualist, integrated realism, see Vetter 2020 for an appeal to affordances to support a non-Humean metaphysics. See also Rydenfelt (2021) for a defense of a Peircean pragmatist hypothetical realism that embeds the judgments of what is representationally real within the goals of scientific practice. I share with his view an acceptance of the contingency of claims of realism. My view also follows Price 2007 in "dimming the light" to reduce the sharp line between realism and irrealism that is required by some forms of representationalism. What I hope to do is to show "how the light gets in" (thank you, Leonard Cohen).

particular experiments and measurement depends on practical and instrumental features of detection. This is different from the language predicting what should be observed according to a theoretical model. That semantic gap must be closed for the results of an experiment to serve as evidence for or against a theory (Darling 2002) and for or against realism.

Additionally, what can be known from experimentation rests on two legs, one on the causal process of intervening on a presumed natural phenomena and the other on the measurement or description of the results of that process. It is not enough to just "twist the lion's tail"[4] to empirically justify a claim about nature. Decisions about what is being twisted, what features are causally engaging with the detecting devise also are required. Most experiments operate on phenomenon-adjacent materials. What is in nature is not what directly interacts in the laboratory. For example, in protein structure prediction, the target protein is no longer culled directly from natural materials, as when cow's blood was used in early studies of the protein hemoglobin, but rather it is made through gene cloning, purification, and isotope-enrichment. The object in laboratory experiments is a form of a protein tuned to be detectible by a specific type of causal interaction.[5] The "same" protein destined for X-ray crystallography detection is prepared very differently than the one for nuclear magnetic resonance spectroscopy experiments. Judgments must be made about how the version of the protein in the causal interactions of an experiment is related to the protein "in the wild."[6] In addition, what counts as causation itself, the criteria for claiming a causal relationship between the phenomenon and instrument yielding data, sits behind the kinds of decisions made about which parts of the initial data output of an experiment are kept and used and which are thrown out and ignored.

The well-known logical problem of underdetermination of theory by data is usually attributed to Duhem, who exposed the role of auxiliary assumptions involved in the designs, practices, and measurements in experiments, all additional to the theory under test, that are needed to derive experimental predictions. Such assumptions are required in the inference from abstract theory to concrete data prediction to specify the experimental set up, the conditions for blocking confounders, etc. Given the logical role

[4] This is attributed to Francis Bacon (perhaps apocryphally) by Thomas Kuhn (1976), to distinguish interventionist experimentation from passive observation.

[5] Acton, Xiao, et al. (2011).

[6] Related judgments are made in the use of model organisms, or representative samples in statistical analysis.

of auxiliary assumptions, when an experimental result does not match the theoretically predicted result, then, Duhem argues, there is no way to isolate the theory under test as responsible for that failure. This is the standard logical account of underdetermination of theory by evidence in the context of theory testing. But Duhem also argued for a form of semantic underdetermination (Duhem 1991; Darling 2002). He identified two components to experimentation. The first he called practical facts, namely what is observed where no physical theory, or at least not the abstract theory to be tested, is required. Practical facts represent the observed results of the causal interaction of the target phenomenon and the detecting device. The second component, Duhem called theoretical facts, namely the predicted observations derived from the theory.

Duhem posed the problem of semantic underdetermination as follows: "The same theoretical fact may correspond to an infinity of practical facts.... The same practical fact may correspond to an infinity of mathematically incompatible theoretical facts." (1991, 152). There is often a mismatch of the precision in a theoretical prediction, especially if it is of a numerical value, and the imprecision accompanying the measurements of the outcome of an experiment. The range of values resulting from repeated measurements, sometimes reported as the average, sometimes "cleaned" by various methods, may be less discriminating than what is deducible from a theoretical model. For example, a theory might predict a precise value, say the melting point of lead as 621.43°F. The experimental protocol may permit a degree of precision only to the whole number. What is the inferential import of the practical measurement, for example, of 621°F to the predicted value of 621.43°F? The practical measurement does not discriminate between 621.00°F, 621.40°F, 621.99°F and so on. Mathematically these are not equivalent, experimentally they are indistinguishable, but the theoretical interpretation has to make a judgment of how to treat them, that is, decide whether they support or refute the prediction of 621.43°F.[7] Duhem argues that requires translation.

> But translation is treacherous: *traduttore, traditore* (to translate is to betray). There is never a complete equivalence between two

[7] See Tal (2017) for an extended discussion of measurement standards and accuracy. On his account, epistemic measurement accuracy is influenced by more than precision. He argues, as I also will, for a pragmatist understanding of accuracy and robustness in which both theoretical and experimental factors are jointly incorporated.

> texts when one is a translated version of the other. Between the concrete facts, as the physicist observes them, and the numerical symbols by which these facts are represented in the calculations of the theorist, there is an extremely great difference. (1991, 133)

The semantic source of underdetermination (Darling 2002) exposes the role of judgments which rely on reasons not contained in either the abstract theoretical model, nor in the concrete causal experiment. These include judgments about what counts as an outlier data point, what counts as noise in the data, what counts as data skewed by systemic "error," or more generally, when variation in the data is taken as the signal and when it is taken as deviation from the signal. Duhem identifies the problem, but what is the solution?[8] What goes into the decisions that permit the empirical results of experimentation to arbitrate between alternative theoretical models? As I will describe below, those judgments appeal to the theory under test, as well as theories of the instruments being used, assumptions about what the bounds are of a phenomenon, and to philosophical frameworks about what counts as, or serves as criteria for, causation generally.

Another difference between abstract theory and concrete data is recognized by Bogen and Woodward's (1988) influential distinction between data and phenomena. According to them, scientific theories predict and explain facts about stable phenomena, like the melting point of lead, or the 3-dimensional structures of a particular protein, or the chemotaxis behavior of e. coli. On their view, data acquired observationally or experimentally can serve as evidence for the existence of such phenomena, but data is not predicted or systematically explained by theories. This is because data are "idiosyncratic to particular experimental contexts, and typically cannot occur outside of those contexts" (pp. 305–6). Data provide a record of what is locally observed, whereas the phenomenon, which is the causal target of a particular experiment generating specific data, goes beyond any single experiment. Phenomena have stable repeatable characteristics that may be detectable by means of a variety of different experiments yielding different data (see also Teller 2010 for a model-theoretic interpretation of their view).

I interpret their distinction in the following terms: natural phenomena are what we posit as the source of the signals entering into causal

[8] See Darling (2003) for different type of realist interpretation of Duhem. See Rydenfelt (2021) for a pragmatist defense of realism.

interactions with detecting devices in experimental, observational contexts (see also Woodward 2000, 2010). The results of experimental causal interactions are represented as data. If the experimental detection devices and procedures are *reliable*, i.e. the produced effects represented as data reflect the properties of the source phenomenon, then, from measurements of the produced signals, we can construct a model of the data that can be used as evidence for the existence of the (stable, non-idiosyncratic) phenomenon. Theories explain and predict features of the phenomenon; experimental data reflect only what is causally detected in an experiment. This is a version of Duhemian semantic underdetermination, cashed out in terms of inferential roles.

Bogen and Woodward (1988) introduced phenomena as a triangulation point between data (which stand in a causal relationship to phenomena) and theories (which inferentially predict and explain phenomena). However, the inferential separation of theories from data (theories don't explain data, they explain phenomena) cannot be strict. Some have argued that theory-ladenness of observation defeats a claim that theory and data are independent (Schindler 2011; see also Boyd and Bogen 2021). Others, like Suppes and Giere, provide more fine grained distinctions among theories that are involved in prediction and experiments that support a more nuanced architecture for the inferential roles of theory and data and thus new spaces to consider the judgments required for translating the results of a particular experiment so that it can speak to a particular abstract theoretical model.

Suppes (1962) articulated a hierarchy of theoretical models[9] involved in assessing the adequacy of causal processes in generating reliable experimental data. On his picture an analysis of a relationship between theory and experiment is iterated at every level of the hierarchy, from non-formalized considerations of confounding factors in the experimental set up, to experimental design choices, statistical models of the data, models of the experiment, and experimental models of the phenomena. Different criteria are applied to determine if an experiment is run correctly, than those invoked in deciding whether the experimental data count as a realization of the theoretical model. In short, Suppes claims that "a whole hierarchy of models stands between the model of the basic theory and the complete

[9] Both Suppes and Giere promoted the semantic view of theories, and the content of scientific theories was interpreted in terms of theoretical models consisting of structures or sets of relations, typically mathematical.

experimental experience. Moreover, for each level of the hierarchy there is a theory in its own right" (260).

What are the relationships among the multiple theories involved in testing the abstract physical theory that represents stable, non-idiosyncratic phenomena? Suppes separates the theory of the phenomenon from the theory of the data in two distinct hierarchical tracks (Figure 4.1). Giere (2010) proposes a hierarchy of levels (Figure 4.2) to locate the two tracks by opposite directions of the (not necessarily strictly inferential) arrows, up from data and down from theory. Yet even with these elaborations, the semantic gap persists. The point of contact between theory and data occurs in the confrontation of measured data from experiment with predicted data from theory. Bogen and Woodward's explanatory difference between phenomena and data, and Suppes's and Giere's hierarchies of theoretical and experimental steps point to important differences. However, the character of the relationship between the distinct components is still in question. We might ask in what ways the theory and the data are independent of each other, and in what ways they are not (see Franklin and Perovich 2021; Epstein and Forber 2013).

I have argued elsewhere (Mitchell and Gronenborn 2017; Mitchell 2020b) that neither strict independence nor a strict hierarchy is adequate to account for the inferential relations among theories and data in scientific practice. Consider the prevalence of semi-empirical modeling where there is

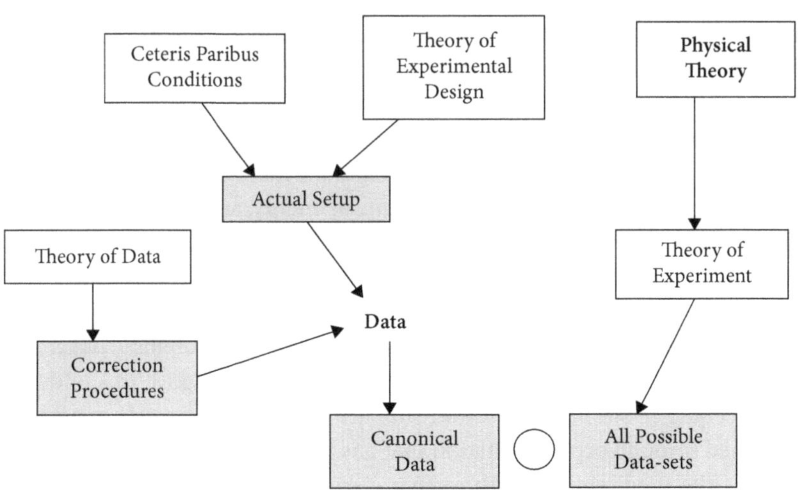

Figure 4.1 Darling's representation of Suppes account (2002, 524).

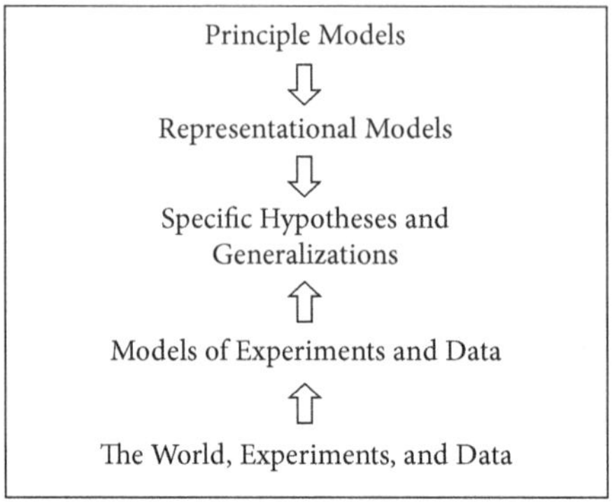

Figure 4.2 From Giere (2010, 270).

extensive feedback between data and theories in generating the point of contact between the two. In those cases, predictions are derived not solely from an interpretation of abstract, formal functions as in, e.g., testing the mathematical structures that provide the content of Newton's laws in different representational models of that structure in falling bodies, springs, or pendula experiments (Giere 1999). In semi-empirical modeling, the formal functions that predict what should be observed in the data are derived from both the theoretical model and from relationships supplied directly by patterns in data. In these cases, the theory–data connection is more complicated. For example, consider the identification of protein structure, i.e. the three dimensional location of the atoms of a protein in its most stable conformation, which is important for understanding protein function. In homologous semi-empirical modeling for predicting protein structure from its amino acid sequence, data is used in a constructive manner. Predictive algorithms embed the results of prior experiments on sequence-similar targets to limit the space of protein conformations that needs to be searched to locate the conformation with the least mean energy, and hence predicted to be the "real," functional atomic structure (Mitchell and Groneborn 2017). "The accuracy of a comparative model is related to the percentage (of) sequence identity on which it is based, correlating with the relationship between the structural and sequence similarity of two proteins" (Baker and Sali 2001, 93). There are a variety of automated search algorithms to determine sequence similarity and fold similarity among the protein structures obtained through experimentation that has been deposited in scientific data banks. The predictive algorithms that start with a comparative homology analysis have, to date, been much more accurate, tested against experimental results for the target proteins, than *ab initio* predictions based only on the sequence of the target protein and molecular dynamics of the physical interactions of the atoms in that protein (for a recent overview of the different approaches, see Deng, Jia, and Zhang 2018; see also Humphreys 1991; Ramsey 1997). A static, hierarchical view of the relationships among theories, predictive models, and experiments obscures some of the processes and judgments required to construct both the empirical and theoretical bases for inferences about protein confirmation. The same is true of inferences in the many and diverse cases in which scientists use semi-empirical models to understand phenomena.

In crafting scientific knowledge, formal theories of the phenomena and data from experiments are inextricably entwined. The theory at test, by its characterization of the phenomena, is required both in designing an experiment and determining how to describe the experimental results that would count as a test of its accuracy. Theories of the experiment and the data are required to set standards of causal reliability so that the appropriate data are considered. The entanglements of theory and experimental observations generate new questions for how realism about phenomena can be warranted. So far, the discussion has been about theories and experiments, both products of human conceptual and causal practices. What else goes into the stronger commitment to the reality of the phenomena that are represented in the confirmed theory and the source of the experimental data?

4.3 Metaphysical Implications of the Complexities of Theory–Phenomena–Data Relationships in Practice

In this section I will consider the sources of warrant for realism about phenomena. There are top-down and bottom-up alternatives to both the content of and the warrant for scientific realism.Anjan Chakravartty (2021) describes what he calls the "realist tightrope," on which one aims to balance between the thinnest and thickest descriptions of what is real, and associates these with bottom-up and top-down methods of warrant. Minimally, a realist stance seeks to anchor the reference to a phenomenon in the world in a way that is independent of individual detections or conceptualizations of it. Maximally, the realist commits to the phenomenon as having precisely the detailed features described in the theoretical model.

These degrees of "thinness" are related to two fundamentalist ways of warranting metaphysical claims about phenomena from scientific practice: a bottom-up approach (entity realism) and a top-down approach (structural realism).[10] The bottom-up approach takes causation as the foundation of positing what's real. Unobservable phenomena are taken to be the cause of experimental data, and thus the interactions and results of empirical practices provide the required metaphysical warrant. Defenders of this view include Ian Hacking (1983) and Nancy Cartwright (2007). Hacking, in discussing an experiment he observed at a Stanford laboratory where electrons and positrons were sprayed, one after the other, onto a superconducting metal sphere, famously claimed that "if you can spray them, then they are real" (1963, 24). Cartwright suggests that what are real are causal capacities in the world, identifying capacities as the source of stable causal laws that we can infer from causal interactions. When a particular experiment yields specific measurements, for Cartwright, the explanation for these measurements posits entities in the world that have the capacity to generate those signals, similar to Bogen and Woodward's stable, non-idiosyncratic phenomena. Entity realism is inferred bottom-up from causal manipulations. Data are evidence for the existence of phenomena.

A fundamentalist top-down view recognizes that unobservable phenomena are the referents of abstract explanatory theories. Structural realists read what is real off of the formal relations represented in scientific models of such theories. Using this strategy, the best confirmed theories are the source of warranting claims about what is real. On this top-down approach what science discovers about nature are not the entities which have causal capacities but rather the structural relations that explain patterns in our observations. The structures described by the mathematical relations in our best theories are taken to be isomorphic or otherwise similar to what is real. John Worrall (1989) defends this type of structural realism when he claims that "On the structural realist view what Newton really discovered are the relationships between phenomena expressed in the mathematical equations of his theory." Current philosophical debates between entity realists like Cartwright (1982) and structural realists like French and Ladyman (2010) exemplify Charkravartty's bottom-up versus top-down fundamentalist dichotomy.

I suggest we break out from the dichotomous choices on offer. I maintain that we need a non-fundamentalist, pragmatist option that jointly uses both strategies for warranting the real. This eschews a static representationalism

[10] See Galison (1988) for a historical analysis of the two sources of fundamentalism.

of the real either as what is presumed to be "in the world" by experimental practice or as referents of theoretical models. What we are warranted in claiming about what is real are not just structures and not just entities, rather it involves the integration of both human interventions and conceptualizations. Neither of the two fundamental strategies alone will capture the judgments required for warranting claims of realism.[11] I suggest that real phenomena are those things in the world that are sufficiently stable to afford the coordination of the results of both causal detection *and* structural representation.

4.4 Affordances, an Interactionist Metaphysics

My view is inspired by the work of the twentieth-century ecological psychologist J. J. Gibson who coined the term "affordance" to explain human and animal behavior (Gibson 2014). Gibson suggested we need a term to identify properties or entities that would convey the joint contribution of both the actor and the environment in which they act. Gibson recognizes the similarity of his proposal to the idea of an ecological niche[12] which identifies the features of the external world that are salient to particular species' activities and actions. Gibson coined the term "affordance" to accommodate that interactive relationship in the psychology of perception. For Gibson, affordances are invariant features of the external environment, that are perceived, classified, etc. dependent on an animal's capacities to interact with it.

"The *affordances* of the environment are what it *offers* the animal, what it *provides* or *furnishes*, either for good or ill. The verb to *afford* is found in the dictionary, but the noun affordance is not. I have made it up. I mean by it something that refers to both the environment and the animal in a way that no existing term does. It implies the complementarity of the animal and the environment (2014, 119)."

[11] Chakravartty's conclusion is to resist the tightrope and embrace a plurality of consistent claims from top-down and bottom-up reasoning regarding what is real. Here I suggest that by identifying the locations of incompatibilities (in data interpretation, theoretical constraints, or philosophical commitments) we can better understand how compatibility is forged in scientific practice.

[12] How to define an ecological niche, and its significance to ecology is still much disputed. See Justus (2019).

Affordances for Gibson are real, objective properties of the environment-plus-organism that make specific behaviors possible. For example, Gibson (2014) indicated that "to be graspable, an object must have opposite surfaces separated by a distance less than the span of the hand" (133). The affordance, however, is not "out there" for the organism to engage, like a disposition or causal capacity. Rather it is constructed by the engagement with the organism.

> An important fact about the affordances of the environment is that they are in a sense objective, real, and physical, unlike values and meanings, which are often supposed to be subjective, phenomenal, and mental. But, actually, an affordance is neither an objective property nor a subjective property; or it is both if you like. An affordance cuts across the dichotomy of subjective-objective and helps us to understand its inadequacy.
>
> (Gibson 2014, 129)

I propose that we consider what is justifiably real in science to have this type of interaction relation. The joint contributions of causally grounded experimental data and theoretically structured representational models together specify what is real and what is not. If we substitute "scientist" for "animal" and "what is real" for "environment" we get a framework that better locates the metaphysical claims warranted by scientific practice. Borrowing some of Gibson's language:

The affordances of real phenomena are what they *offer* the scientist, what they *provide* or *furnish to experiment and representation*. I (Mitchell) mean by it something that refers to both the causal properties in nature and the representational framework of the scientist. It implies the complementarity of both in establishing what is real.

This interactionist approach to realism rejects both top-down and bottom-up fundamentalism. There is no fixed ontology of entities or fixed ontology of structures. Instead, what we project back as being constitutive of nature is contingent on a complementary relationship between what we can detect and how we represent it.[13] Nature is independent of us. The universe and our planet existed prior to humans and probably will exist post the human era. But what is viable as a metaphysics of what is **real** according to our science is what we, as limited beings, are justified in claiming about nature. All we can justifiably say yields a contingent, pragmatist ontology of

[13] See Chirimuuta (2017) for an analogous argument about color vision.

stable, detectable and representable, entities and relations, i.e. affordances built from both experiment and theory.

Real affordances can be represented as entities with stable dispositional causal properties that we engage in experimental practices. Real affordances can also be represented as stable structures described in the relations constituting our best theories. Some form of stability is a minimal metaphysical presumption for doing science. In what follows I will investigate the kinds and character of stability required by experiment and model. Specifically, I will address a key question for a pragmatist form of realism: What are the roles of causes and concepts in constructing the real?

4.5 Reliability of Data, Theory, and Frameworks

I have suggested that affordance metaphysics exposes the interactive roles of theories, conceptual frameworks, and causal-experimental models in warranting metaphysical inferences. An assumption of any experimental detection procedure is that there is a phenomenon,[14] independent of the detecting device, that is being causally engaged. Real phenomena afford stable, repeatable causal interactions. They might be detected by means of a variety of different procedures, each producing idiosyncratic data as emphasized in Bogen and Woodward's distinction. The agreement between a theoretical model-based prediction from phenomena to data and the experimental inference from data to phenomena are the joint source of justification for the existence of the phenomenon and relations specified in the model. If alternative models predict divergent data, and if the experiments are sufficiently precise to distinguish between the competing models, then the data provides the kind of contrastive confirmation that justifies accepting one model over another as a more accurate representation of what is real in nature. If the data of multiple, diverse experiments all agree with the model-based prediction, then that is taken as the strongest evidence that the multiply detected phenomenon is real (Salmon 1985). In what follows, I will dig deeper into the structure of assumptions that ground judgments of realism from the convergence of experimental results.[15]

[14] By phenomena I follow Bogen and Woodward (1988) to mean the stable relations or entities that are not directly observable which are candidates for what is real.

[15] In Mitchell (2020a, 2020b), I have presented arguments for how divergent data from experiments can also be a route to increased representational accuracy.

When phenomenal stability is conjoined with reliable detection, then convergent data are taken to support claims of realism. Causal reliability is a judgment about the causal process generating data in an experiment which contributes to the inference from data to phenomenon for each type of experiment.[16] When multiple different types of experiment generate data from which "the same" phenomenon is inferred, then realism is the conclusion of a no-miracles argument. When there are substantially different, if not strictly independent, ways of causally detecting the source of the signals in the experiments, and they all yield the same measurement, then, as the no miracles argument goes, it must be because the source is a real phenomenon independent of each individual experiment with its own idiosyncratic assumptions or models of the experiment. Experimental convergence is taken as the strongest "bottom-up" evidence for realism.

What is required for this argument to support a claim of realism? First, the causal processes (phenomenon to data) realized in the experiments should, themselves, be reliable. Second the multiple experiments should be independent of each other. I will discuss first the issue of independence, then turn to reliability.

4.5.1 Independence of Experiments

There are three types of independence that are typically invoked in arguments for convergence or robustness of experimental results to ground a claim of realism. First, the predictions from the theory should be independent of the actual results obtained by running the experiment. Second, the experimental design and inference to the measured data in a particular experiment should be independent of the theory/model under test. Third, the multiple experiments whose results converge, should be independent of each other.

Clearly there are theories involved in designing and performing an experiment. The models of the experiments presuppose theories in order to set up the conditions in which the phenomenon would be detectable by the instruments, as Duhem emphasized. A central concern has been which conceptual and theoretical assumptions are required for the data from a

[16] See also Woodward (2010) for a description of reliability in inferences from statistical evidence.

single or multiple experiments to be a test of a prediction about the phenomenon. Suspicion is raised about whether the data from an experiment can serve as a test of a theory if, for example, the predictions are built after the fact from the observed data. Overfitting parameters or tweaking a model after the experiment has been run can be epistemically perilous (see Epstein and Forber 2013). But it is also the case that data about how the experimental apparatus operates, which may be gleaned from results of running experiments, is required in order to generate a prediction from the theory in the first place. Which initial conditions need to be in place, which causes can be blocked from confounding the results? For an experiment to be "about" the phenomena described in a theory and the prediction to be "about" the experimental results, some coordination is required. To produce epistemically relevant information, the Duhemian semantic underdetermination relationship between theoretical predicted facts and practical experimental facts requires a connection between them be made. To do so undermines a view of the strict independence of data from theory for it to serve as evidence for that theory.

Second, instruments selectively interact with some features of the target phenomena, not all features, and in this sense, are causally biased (see Giere 2006). Judgments of convergence and its associated warrant for realism require a prior decision that the multiple experiments are of the same phenomenon. Measurements or some other representation of the causal effects of the instrument/phenomenon interaction encode conceptual and theoretical assumptions about the salience of the feature targeted in an experiment for the theoretical fact about the phenomenon. For example, experimental determination of protein structure is predominantly achieved by X-ray crystallography or nuclear magnetic resonance spectroscopy. X-ray experiments target the electron clouds of crystalline atoms in a protein which diffract a beam of incident X-rays. The angles and intensities of the diffracted beams are measured and a three-dimensional density map of the electrons produced. From this, the relative positions of the atoms in the crystallized protein can be determined. In contrast, NMR protein experiments target quantum mechanical properties of the nuclei of the atoms (nuclear spin polarization) in the protein. A protein is placed inside a large magnet, radio frequency signals are sent through the sample, and the absorption of those signals is measured which allows the relative distances between atoms in the protein sample to be inferred. When the two different experimental protocols agree in their inferred experimental model of the protein structure, this is taken to be evidence that the real structure has been detected.[17] The behavior of electrons and the behavior of nuclei are related by theory, which permits us to say that the two experimental methods are "about" the same target phenomenon, namely the atomic structure of the same protein, even though the specific causal interactions in the experiment involve very different affordances—the interaction of atomic *electrons* with monochromatic X-rays and the interaction of atomic *nuclei* with certain kind of variation in a strong magnetic field. The affordances are constructed jointly from the theoretically "same" external target (atoms in a protein) *and* the different experimental environments and detecting devices. The two different affordances produce two different, theoretically linked, and reliable views of realities of protein structure.

Tal argues that: "To attain the status of an outcome, a set of values must be abstracted away from its concrete method of production and pertain to some quantity objectively, namely be attributable to the measured object rather than the idiosyncrasies of the measuring instrument, environment, and human operators." (2017, 35). But what is it for the quantity to be objective? On my view, it is for there to be *integrated affordances*. I have argued elsewhere (Mitchell 2020a, 2020b) that multisensory perception provides an informative analogy. When more than one sense modality (vision and hearing) uptakes information from the environment, a signature consequence of their neuronal integration is the superadditivity of their effects. That is, the number of neurons firing in multisensory integration is more than the sum of the firing rate of the two individual modalities. Superadditivity also is seen in multi-modal integration in time-to-task completion studies (e.g., grasping an object). Like Gibsonian affordances, multisensory integration processes are context and modality dependent. Vision is more reliable for spatial representation while audition is more reliable for temporal representation. In specific contexts, visual input will dominate over auditory input; in other contexts, the reverse ordering will occur. Yet, integrating the two provides superadditive accuracy of representation (Van Atteveldt et al. 2014). Just as brains integrate input from the modalities by which they acquire information, selectively targeting different types of signals in ways that permit more accurate representations and more

[17] The story is much more complicated than this, of course. Each experimental protocol generates an ensemble of conformational structures that are consistent with their data, not a single structure. So agreement has to do with the overlap of the ensembles.

effective action, so too, I argue, does the integration of the multiple causal, theoretical and conceptual "modalities" of scientific practice.

Theories are essential for both performing an experiment, in the form of a causal theory of the experiment, and for identifying when two convergent (or divergent) experiments are about the same phenomenon. Strict independence of theory and experiment is not possible. Does that mean we have to accept an all embracing Quinean holism where every scientific claim is connected to every other scientific claim? In his seminal work of the history of the thermometer, Hasok Chang (2004) argues that scientists can avoid the most extreme forms of holism by adopting a "principle of minimalist overdetermination." Chang means by overdetermination the type of convergence discussed above, namely, agreement by different methods on the measurement value ascribed to some phenomenon, e.g., a temperature determined by both theoretical inference calculation and measurement by a mercury thermometer or measured by two different types of thermometer. He agrees that invoking auxiliary hypotheses is necessary in order to make predictions, build apparatus and interpret the results of an experiment, but by minimizing those assumptions, Chang argues, the damage can be contained. "The heart of minimalism is the removal of all possible extraneous (or auxiliary) non-observational hypotheses" (Chang 2004, 94). His approach suggests we should prefer the most direct, or least theory-mediated inference, between the causal experimental interaction and the representational measurements.

One might interpret "minimalist" to require that the theories of the experiment should rely on *as few assumptions* about the function that associates the target feature with the measurement as possible. By minimizing the number of auxiliary assumptions Chang's thesis would suggest we can provide the strongest grounds for taking experimental results to be confirmations or refutations of the hypothesis being tested. The theory of the causal interaction between phenomena and instrument should, if this interpretation of Chang is correct, rely on no more assumptions implicated by the theory under test than is necessary. This approach has the virtue of recognizing the ineliminabilty of influence of theoretical assumptions in acquiring measurements in an experimental set up while attempting to avoid ad hoc, unreliable confirmation. However, this interpretation implies that using the fewest number of theoretical assumptions in producing data [minimization] is what matters. Given Chang's extensive appeal to scientific practice and his recent pragmatist account of realism (2022), it is clear that this interpretation is not right. I suggest, and I suspect Chang would agree, that what matters most is *what* is assumed in an experimental set up, not just *how many* assumptions are used. By specifying what parts of the theory under test are assumed, the decision that convergence of multiple experiments has been achieved and realism can be justified can be made openly and reflectively.

4.5.2 Experimental Reliability in Data-to-Phenomena Inference

I have suggested that affordances are what science can justify as being *real*. Neither causally detectable phenomena alone nor theoretical structures alone are sufficient to characterize the causal interactions with multiple detection devices that produce data which support claims of realism. Such claims always involve a translation from Duhemian practical facts, or minimal descriptions, of the upshot of the causal process enacted in an experiment into Duhemian theoretical facts which support or fail to support theories describing the relations that hold among real phenomena. There are assumptions and decisions throughout this chain of inferences that shape the claims of what is real and depend on both our goals and our epistemic and technical limitations. I now will turn to the assumptions that enter into this inferential process through judgments made about the reliability of data.

For experiments to provide grounds for positing something about unobservable phenomena, we need to assume that the data is reliably reflecting the causal input from the target phenomena. There are important issues about how we reason about eliminating random and systematic influences in data production to filter out the non-target contributions to the initial data. One role for multiple experiments is to expose the systematic, non-target contributions of the individual experimental protocols and operations (Kuorikoski and Marchionni 2016; Mitchell 2020b). Random and systematic sources of non-target signal require different treatment in reaching the decision that the "cleaned" data provides information about the target phenomenon. Suppes (1962) identified descriptive features of the data that could provide clues about the target and non-target sources. Random noise is identified by characteristic instability in the data over multiple experimental runs (Steinle 2002) and its effects can often be averaged out. Experimentalists compare iterations of experiments to distinguish random and systematic sources of data (e.g., Morris 1992). Systemic, non-target effects are not simply averaged, but may be managed in a variety of ways including redesigning experimental protocols to block non-target causes, to re-interpreting measurements to adjust for known biases in the instruments.[18] Experimental repetition and comparison can expose features of data that can be associated with random and systemic, non-target, contributions to the measured signals. But to serve this function, both theoretical assumptions about what *kind* of phenomenon is being tested and conceptual assumptions about causality itself are involved.[19]

Here I will examine the roles played by the theoretical characterization of the phenomena and the philosophical theory of causation in judgments about the reliability of experimentally produced data. Both are essential to the justification of claims of realism. If there is a real phenomenon, there will be some form of stability across experimental detection. And if the detection process and measured data are causally and inferentially reliable, then convergent data can be taken as justification of realism. But, I will argue, the judgment of the reliability of detection depends upon both the kind of phenomena investigated and the type of causation invoked.

Woodward's (2005) influential account of causation characterizes causal stability as invariance under intervention. On his view, invariance need not be exceptionless (see also Mitchell 1997, 2000). To count as causal, "A generalization can be invariant within a certain domain even though it has exceptions outside that domain. Moreover, unlike strict lawfulness, invariance comes in gradations or degrees" (Woodward 2000, 199). Woodward's interventionist account of causation is not intended to be merely methodological, but is a philosophical, or conceptual account of causation: "...for Y to change under an appropriate intervention on X just is what it is for X to cause Y" (2000, 204–5, fn1).[20] One alternative account of causation that Woodward rejects is a regularity account that requires strict laws. On the strict law view, for X to cause Y is for the generalization describing the causal relationship to be universal, exceptionless, true, and naturally

[18] How wonderful that some of the programs for cleaning data in structural biology Xray crystallography experiments are named POINTLESS and AIMLESS. See Evans and Murshudov (2013) for a serious discussion of the criteria for deciding the `resolution' of a measured data set.

[19] The theoretical differences between Quetelet and Galton, for example, treated variation in the values of the measured data of a population in completely opposite ways.Quetelet, by appealing to the mean-squared law of astronomers, assumed there was a "real" singular height of Scottish soldiers and the measured variation could be averaged away to find it. Galton, in contrast, assumed the variation was the real phenomenon, and the average was a statistical description of the data—not the world. See Stigler (1986).

[20] Woodward's most recent work has explicated the psychological foundations for his counterfactual account of what humans mean by causation (Woodward 2021).

necessary. There are other philosophical theories of causation including probabilistic, primitivist, eliminativist, and process accounts (Gallow 2022; Frisch 2022; Hitchcock 2021). For the purposes of this chapter, I will consider only interventionism and strict lawful regularity accounts.

Causal reliability in experimentation is a judgment about how well the causal signal from the phenomenon is reflected in the data of the detecting device, and hence how well it supports the reliability of the inferences from data to phenomena and subsequently in support of theoretical models. Sometimes reliability is parsed in terms of validity. Campbell (1957) introduced the distinction between internal and external validity and further elaborations have been developed, but basically experimental validity refers to the accuracy of an experimental measurement. Validity has been identified with replicability of results and applicability to targeted real-world situations. For example, Guala characterizes internal and external validity as follows: "the result of an experiment E is internally valid if the experimenter attributes the production of an effect B to a factor (or set of factors) A, and A really is the (or a) cause of B in E. Furthermore, it is externally valid if A causes B not only in E, but also in a set of other circumstances of interest, F, G, H, etc." (Guala 2003, 1198). But the "really is" language begs the question I want to explore, namely, how can we justify claims of what things "really are" on the basis of the reliability of experiments? We incur a problematic circularity if reliability requires we know in advance that reality for which the experimental practice is designed to provide evidence.

I argue that the affordances constructed from philosophical accounts of causation, theories of the phenomenon, and judgments of experimental reliability support a non-foundationalism in claims about realism, hence making room for the interactionist affordance alternative I propose.[21]

On Woodward's account of causation, causal relations and the generalizations that describe them can differ in degrees of invariance and still count

[21] Reiss 2019 argues against both the evidential reasoning built on the distinction between internal and external validity and methodological foundationalism. He claims the latter assumes some forms of experimentation are intrinsically reliable, as seems to be the case in medicine's overwhelming endorsement of the "gold standard" of randomized controlled studies as exclusively justificatory. Reiss suggests this form of methodological foundationalism can be seen as an attempt to realize the ideal interventionist logic of Woodward's invariance account of causation. Reiss suggests that "Reasoning concerning target systems should begin with a hypothesis about that system and ask what types of evidence we need to establish that hypothesis." Reiss (2019). I share Reiss's conclusion, but arrive at it by a different route and deploy it for different purposes.

as causes. Whether data counts as reliable evidence of a real phenomenon or not depends on the kind and degree of stability or invariance that is attributed to the causal structure by a theory about the phenomenon. Contrary to the claim that the theory of the phenomenon is or should be independent of the experimental data, for data to be deemed reliable, that theory has to be invoked.

The roles of the theory of the phenomena and causal theory in determining reliability is somewhat invisible in the easy cases of the most stable phenomena, like the melting point of lead, discussed in Bogen and Woodward (1988). In what follows I will illustrate the roles in a case of emergent biological phenomena.

What is the relationship between the reliability of an experiment and the degree of invariance of the causal relation? The melting point of lead has the properties that fit the Bogen-Woodward conception of phenomena, i.e. stability over a wide range of experimentally enacted causal conditions. On Woodward's interventionist account (2005) of what causality is, that means the functional relation among the variables explicitly represented in the theoretical model predicting the melting point is *invariant under intervention*. For Woodward's account, the functional relation is also independent of other causal factors operating, i.e. it satisfies the condition of *modularity* or independent disruptability. And the functional relation is *insensitive* to a host of conditions of varying, non-represented or exogenous background conditions. The melting of lead is predicted to occur when the forces associated with the thermal motion of the free electrons exceed the electromagnetic forces holding the electrons and nuclei in a solid lattice structure or, as Bogen and Woodward put it, "the melting of lead occurs whenever samples of lead are present at the appropriate temperature and pressure, and results from a characteristic change in the crystalline structure of this metal" (1988, 319). There are *ab initio* theoretical predictions for melting points of metals[22] and there are experimental measuring procedures. While experimental measures deliver a range of data points, their average is of 327.5 degrees Celsius, though no single measurement may precisely match that value. When does the relatively close agreement between the theoretically predicted value and data obtained from multiple different measuring techniques warrant belief in the reality of the phenomenon?

[22] For an example of an ab initio prediction of the melting temperature of a metal see: "Thermodynamic calculation of the melting point of aluminum" (Belan-Gaiko et al. 1980).

For the inference to realism from convergence to be justified, the range of variation of the measured values within and between experiments has to be sufficiently narrow for the practical facts of the experiment to be interpreted as satisfying the theoretical fact of the prediction. But what if a new type of measuring technique is developed that in multiple replications indicates a wildly different value? Based on the past stability of measurements we might infer that the new protocol is not valid. Or, if there is theoretical justification for the new technique, it might explain why the value is far from the older techniques by identifying some systematic bias shared by all the older techniques. We might accept the new results and revise the account of the phenomenon based on it.

On the strict lawful regularity account of causation, science cannot tolerate inconsistent results. This account of causality requires a strict reading of Bogen and Woodward's 1988 claim the phenomenon will occur "*whenever* samples of lead are present at the appropriate temperature and pressure." Of course one could question whether the "appropriate" conditions are met. But presuming a shared judgment of what is appropriate, the strict laws view of causation would demand there be no exceptions. However, Woodward's interventionist account permits causation with degrees of invariance under intervention, rather than all or nothing strictness. If the conditions of the new experiment push the phenomenon outside the range of its causal invariance, both the new and old measurements can be deemed accurate, and hence both experimental processes would be causally reliable. As I will argue below, to decide if an experimental result inconsistent with the theoretical prediction is reliable (and hence a refutation for the strict law view of causality) or reliable but merely reflecting the boundaries of the range of invariance (not necessarily a refutation on the invariance view of causality) or not reliable at all, requires a theory about the type or features of the phenomenon being investigated.

The significance of theory in determining reliability of experiments and inference to the reality of the phenomenon becomes clearer when we consider experimentation on dynamically complex phenomena.Consider cases where a system-level property is caused by behaviors of the components of that system. Lead melts at a temperature (its melting point) when the thermal motions of the free electrons in the atoms composing the lead breaks the electromagnetic forces holding the electrons and nuclei in a solid lattice. There may be exceptions, i.e. lead fails to melt at its "melting point," even when the functional relationship between the variables for thermal motion and electromagnetic forces satisfy the theory, but this case is pretty close to a strict generalization.[23]

Complex systems phenomena, like the genotype–phenotype relationship where a gene is taken to be the cause of the trait (e.g., "gene for" language in the gene for Huntingdon's chorea, or the gene for melanism in peppered moths, etc.), are less stably realized, more fraught due to exceptions, than melting points of metals. Depending on the account of causation, and the theory of the phenomenon, the reliability of experimental detection of the phenomenon will be judged differently. This is vividly displayed in the history of gene knock-out experiments (see Mitchell 2009) where intervening to remove (or silence) particular genes in an organism is used to identify their phenotypic function or effect.[24] Comparing the phenotypes of organisms with, and without, a particular gene follows the logic of a controlled experiment. Intervene on only one gene, leaving the rest of the genome intact, and any changes to the phenotype should expose the causal effect of that gene. In roughly 30 percent of experiments with viable knockout mutants, there is little or no phenotypic difference. For some, including Mario Capechi who invented the knock-out technique, the inference from these "failed" experiments was that the experiment was not done correctly and, therefore, the results are not reliable. For others, the best inference was that if the target gene has little phenotypic effect in a relatively high percentage of cases, it should not be identified as the cause of the phenotype. A third alternative was proposed (Greenspan 2001) that is based on the theory that there is a robust dynamics of the network of genes that interact in the causal production of a phenotype. On Greenspan's account, when one gene is silenced, there is a reorganization of the network of genes such that, in many of knockout mutants, the normal phenotype is still produced, but by a different causal pathway. This is not a case of redundancy, where there are back up copies of genes that step in when one token is removed, but a case of "degeneracy" (Edelman and Gally 2001) where new causal networks of

[23] Of course, there is the standard move in the strict law camp to introduce ceteris paribus conditions to preserve universal exceptionless. The interventionist might appeal to something similar in determining what are the "appropriate" conditions, but if interventionism is to be an alternative that permits "degrees of invariance" in contrast to strictness it must permit some attribution of causation in cases where the strict view would not. See also Woodward (2002) and Mitchell (2002).

[24] Scientific techniques have advanced in precision since knock-out experiments with the development of CRSPR but the issues of the warrant for reliability and realism occur with all experiments.

interaction arise in the absence of the knocked-out node. "The relationships that have been described as pathways are no doubt real, but they need not be invariant. Their relationships are embedded in broader and more plastic networks that can be reconfigured depending on the immediate circumstances" (Greenspan 2001, 386).

If data from different experiments converge, then the inference is that the experiments are reliable and the phenomenon is real. But if data from different experiments diverge, then, as in the case of knockout experiments, what is inferred exposes how judgments about reliability and reality are shaped by theories of the phenomena and philosophical accounts of causation.

Woodward (2010) further elaborated his analysis of causation broadening the scope of judgments that can be made about causation beyond a dichotomous framework in which something either is a cause or is not a cause. He suggests that instead there are important distinctions among causes with respect to degrees of stability, level of causal description, and causal specificity drawing on examples of genetic causation. I have also argued that a more nuanced account of causal dependence (Mitchell 1997, 2000, 2009) better captures scientific practice and permits us to see the similarities and differences among causal claims made in different sciences. Here, I want to emphasize how a more nuanced view exposes the interactive roles of multiple assumptions in making causal judgments, and thus in shaping and justifying claims of realism.

The reliability of knockout experiments is indicated by but not constituted by their replicability. What counts as a successful or failed replication depends on both a theory of the type of phenomenon being tested and the causal features and pathways that investigators associate with experimental interventions. Are traits caused by individual genes, or feedback sensitive reorganizing networks of genes? Are causes maximally stable, expressing universal laws or are causal relationships stable only in some domains of invariance, but unstable outside of those domains? If a knockout experiment to determine the genetic cause of a phenotypic trait does not show a difference between organisms with or without that gene, or if multiple knockout experiments diverge, then inferences to what is real will track the answers to these questions. On a strict regularity laws view of causation, any failure of stability, no effect, or diverging effects entail that either one or more experiments is unreliable, or there is no real causal phenomenon. These were both responses to the surprising result that 30 percent of knockout experiments fail to display a stable gene–phenotype relationship. On an invariance under intervention view of causation, in the same situation, one could infer that one or more experiments was unreliable, if the expectation of the theory of the phenomenon was for invariance under the conditions of the experiment. But one could legitimately infer that either the conditions of the experiment were outside the domain of invariance (the experiment was reliable) or the phenomenon was complex and robust and hence its stability would not be captured by the gene knockout experiments (the experiment was reliable but not for that type of phenomenon).

The claim that a phenomenon is real acquires empirical warrant from the reliable replication of a single type of experiment and even more so from reliable reproduction of results from diverse types of experiments. Underlying this warrant are judgments about the reliability of the experiments to causally engage certain features of the phenomenon which it targets. Instrumentally "observed" causal processes are perspectival, detecting only those invariant affordances which make a difference to the actions of detection devices. To go beyond the Duhemian practical facts that reliable causal interactions produce, in order to attribute reality to more than local, idiosyncratic results requires invoking abstract, theoretical representations and abstract structures, i.e. theories of the phenomenon (simple, complex, robust, etc.). Only then can multiple experiments of the "same" phenomenon accrue the added-value support of convergence inferences for realism. Clearly, what is permitted, condoned, or forbidden by a concept of and criteria for causality itself will also guide judgments of reliability and hence of realism. To understand why there is a scientific disagreement about whether or not a purported phenomenon is real or not, we need to investigate all three contributing factors in establishing the real affordances of nature—experimental data, theoretical representations, and the causal framework. Real phenomena, external to an observer, afford detection by means of their causal interactions as characterized by theories of the phenomena—biased or perspectival though they be—and afford representation by functional or structural theories of causal relations, partial and conceptually bound though they be. As experiences, experimental techniques, conceptual re-orientations, and theoretical innovations change, our warrant for claiming which phenomena are real will track those changes.

4.6 Conclusion

What justifies a claim of realism in science is a function of the coordination of experimental data (causal reliability), theoretical expectation (types of phenomena with differing degrees or types of stability), and a theory of causation (strict laws, invariance, etc.). What is real is an *affordance whose identification is built from the coordination of reliable interaction with what is external to us and the theoretical and philosophical frameworks constraining and structuring what we can represent.* Ian Hacking (1992) argued for anti-realism by appeal to a similar coordination claiming that "Stable laboratory science arises when theories and laboratory equipment evolve in such a way that they match each other and are mutually self-vindicating" (1992, 56) but that when that happens "we have not read the truth of the world" (58). Rather, "There usually were not some preexisting phenomena that experiment reported. It made them. There was not some previously organized correspondence between theory and reality that was confirmed...The process of modifying the workings of instruments—both materially...and intellectually...furnishes the glue that keeps our intellectual and material world together. It is what stabilizes science" (Hacking 1992, 58).

I suggest that the bearable thinness of being permits a pragmatist realism that reflects the role of human judgments about reliability, types of phenomena and causation in identifying the real affordances that nature permits. This non-fundamentalist, pragmatist approach to realism embraces the contingency of judgments about what is real that depend on what makes a difference both for causal detectability (bottom-up) and structural representability (top-down). This is the kind of realism our scientific practices can bear.

Acknowledgments

I would like to thank Holly Andersen, Joel Smith, and Dana Matthiessen for extremely helpful feedback on this chapter, and all those who participated in the PragMaps workshops for stimulating discussions.

References

Acton, T. B., Xiao, R., Anderson, S., Aramini, J., Buchwald, W. A., Ciccosanti, C.,... & Montelione, G. T. (2011). Preparation of protein samples for NMR structure, function, and small-molecule screening studies. In *Methods in enzymology* (Vol. 493, pp. 21–60). Academic Press.

Andersen, H. K., & Mitchell, S. D. (2023). Pragmatism for Philosophy of Science. (this volume).

Baker, D., & Sali, A. (2001). Protein structure prediction and structural genomics. *Science, 294*(5540), 93–6.

Belan-Gaiko, L. V., Bogdanov, V. I., & Fuks, D. L. (1980). Thermodynamic calculation of the melting point of aluminum. *Soviet Physics Journal, 23*(12), 1009–11.

Bogen, J., & Woodward, J. (1988). Saving the phenomena. *The philosophical review, 97*(3), 303–52.

Boyd, Nora Mills, and Bogen, James (2021). Theory and observation in science, In Edward N. Zalta, ed., *The Stanford encyclopedia of philosophy* (Winter 2021 Edition). https://plato.stanford.edu/archives/win2021/entries/science-theory-observation/.

Campbell, D. T. (1957). Factors relevant to the validity of experiments in social settings. *Psychological bulletin, 54*(4), 297.

Cartwright, N. (1982). When explanation leads to inference. *Philosophical Topics, 13*(1), 111–21.

Cartwright, N. (2007). *Hunting causes and using them: Approaches in philosophy and economics*. Cambridge University Press.

Chakravartty, A. (2021). Realist representations of particles: The standard model, top-down and bottom-up. In T. D. Lyons & P. Vickers, eds., *Contemporary scientific realism: The challenge from the history of science* (pp. 350–73). Oxford University Press.

Chang, H. (2004). *Inventing temperature: Measurement and scientific progress.* Oxford University Press.

Chang, H. (2022). *Realism for realistic people: A new pragmatist philosophy of science*. Cambridge: Cambridge University Press.

Chirimuuta, M. (2017). *Outside color: Perceptual science and the puzzle of color in philosophy*. MIT Press.

Darling, K. M. (2002). The complete Duhemian underdetermination argument: scientific language and practice. *Studies in History and Philosophy of Science Part A, 33*(3), 511–33.

Darling, K. M. (2003). Motivational realism: The natural classification for Pierre Duhem. *Philosophy of Science*, 70(5), 1125–36.

Deng, H., Jia, Y., & Zhang, Y. (2018). Protein structure prediction. *International Journal of Modern Physics B, 32*(18), 1840009.

Duhem, P. M. M. (1991). *The aim and structure of physical theory* (Vol. 13). Princeton University Press.

Edelman, G. M., & Gally, J. A. (2001). Degeneracy and complexity in biological systems. *Proceedings of the National Academy of Sciences*, *98*(24), 13763–8.

Epstein, B., & Forber, P. (2013). The perils of tweaking: how to use macrodata to set parameters in complex simulation models. *Synthese*, *190*(2), 203–18.

Evans, P. R., & Murshudov, G. N. (2013). How good are my data and what is the resolution?. *Acta Crystallographica Section D: Biological Crystallography*, *69*(7), 1204–14.

Franklin, A. and Perovic, S. (2021). Experiment in physics. In E. N. Zalta, ed., *The Stanford encyclopedia of philosophy* (Summer 2021 Edition). https://plato.stanford.edu/archives/sum2021/entries/physics-experiment/.

French, S., & Ladyman, J. (2010). In defence of ontic structural realism. In *Scientific structuralism* (pp. 25–42). Springer, Dordrecht.

Frisch, Mathias (2022). Causation in physics. In Edward N. Zalta, ed., *The Stanford encyclopedia of philosophy* (Spring 2022 Edition). https://plato.stanford.edu/archives/spr2022/entries/causation-physics/.

Galison, P. (1988). History, philosophy, and the central metaphor. *Science in Context*, *2*(1), 197–212.

Gallow, J. Dmitri (2022). The metaphysics of causation. In Edward N. Zalta, ed., *The Stanford encyclopedia of philosophy* (Summer 2022 Edition). https://plato.stanford.edu/archives/sum2022/entries/causation-metaphysics/.

Gibson, J. J. (2014). *The ecological approach to visual perception: classic edition.* Psychology Press. First published in 1979 by Houghton, Mifflin and Company.

Giere, R. N. (1999). *Science without laws.* University of Chicago Press.

Giere, R. N. (2006). *Scientific perspectivism.* Chicago and London: The University of Chicago Press.

Giere, R. N. (2010). An agent-based conception of models and scientific representation. *Synthese*, *172*(2), 269–81.

Greenspan, R. J. (2001). The flexible genome. *Nature Reviews Genetics*, *2*(5), 383–7.

Guala, F. (2003). Experimental localism and external validity. *Philosophy of science*, *70*(5), 1195–205.

Hacking, I. (1983). *Representing and intervening: Introductory topics in the philosophy of natural science.* Cambridge University Press.

Hacking, I. (1992). The self-vindication of the laboratory sciences. In A. Pickering (ed.), *Science as Practice and Culture.* University of Chicago Press. pp. 29–64.

Hitchcock, Christopher. (2021). Probabilistic causation. In Edward N. Zalta, ed., *The Stanford Encyclopedia of Philosophy* (Spring 2021 Edition). https://plato.stanford.edu/archives/spr2021/entries/causation-probabilistic/.

Humphreys, P. (1991). Computer simulations. In A. Fine, M. Forbes, & L. Wessels, eds., PSA 1990, vol. 2 (, pp. 497–506). Philosophy of Science Association.

Justus, J. (2019). Ecological theory and the superfluous niche. *Philosophical Topics*, *47*(1), 105–24.

Kuhn, T. S. (1976). Mathematical vs. experimental traditions in the development of physical science. *The Journal of Interdisciplinary History*, *7*(1), 1–31.

Kuorikoski, J., & Marchionni, C. (2016). Evidential diversity and the triangulation of phenomena. *Philosophy of Science*, *83*(2), 227–47.

Mitchell, S. D. (1997). Pragmatic laws. *Philosophy of science*, *64*, S468–S479.

Mitchell, S. D. (2000). Dimensions of scientific law. *Philosophy of Science*, *67*(2), 242–65.

Mitchell, S. D. (2002). Ceteris paribus—an inadequate representation for biological contingency. *Erkenntnis*, *57*(3), 329–50.

Mitchell, S. D. (2003). *Biological complexity and integrative pluralism*. Cambridge University Press.

Mitchell, S. D. (2009). Unsimple truths. In *Unsimple Truths*. University of Chicago Press.

Mitchell, S. D., & Gronenborn, A. M. (2017). After fifty years, why are protein X-ray crystallographers still in business?. *The British Journal for the Philosophy of Science*, 68(3), 703–23.

Mitchell, S. D. (2020a). Through the fractured looking glass. *Philosophy of Science*, *87*(5), 771–92.

Mitchell, S. D. (2020b). Perspectives, representation, and integration. In M. Massimi & C. D. McCoy (eds.), *Understanding perspectivism: Scientific challenges and methodological prospects*. Taylor & Francis, pp. 178–93.

Morris, G. A. (1992). Systematic sources of signal irreproducibility and t1 noise in high-field NMR spectrometers. *Journal of Magnetic Resonance (1969)*, *100*(2), 316–28.

Price, H. (2007). Quining naturalism. *The Journal of Philosophy*, *104*(8), 375–402.

Ramsey J. L. (1997). Between the fundamental and the phenomenological: The challenge of "semi-empirical" methods. *Philosophy of Science*, *64*, 627–53.

Reiss, J. (2019). Against external validity. *Synthese*, *196*(8), 3103–21.

Rydenfelt, H. (2021). Realism without representationalism. *Synthese*, *198*(4), 2901–18.

Salmon, W. C. (1985). Scientific explanation and the causal structure of the world. In *Scientific explanation and the causal structure of the world*. Princeton University Press.

Schindler, S. (2011). Bogen and Woodward's data-phenomena distinction, forms of theory-ladenness, and the reliability of data. *Synthese, 182*(1), 39–55.

Stigler, S. M. (1986). *The history of statistics: The measurement of uncertainty before 1900.* Harvard University Press.

Suppes, Patrick (1962). Models of data. In E. Nagel, P. Suppes, and A. Tarski, eds., *Logic, methodology and philosophy of science: Proceedings of the 1960 International Congress* (pp. 252–61). Stanford University Press,.

Steinle, F. (2002). Experiments in history and philosophy of science. *Perspectives on Science, 10*(4), 408–32.

Tal, E. (2017). Calibration: Modelling the measurement process. *Studies in History and Philosophy of Science Part A, 65*, 33–45.

Teller, P. (2010). "Saving the phenomena" today. *Philosophy of Science, 77*(5), 815–26.

Van Atteveldt, N., Murray, M. M., Thut, G., & Schroeder, C. E. (2014). Multisensory integration: flexible use of general operations. *Neuron, 81*(6), 1240–53.

Vetter, B. (2020). Perceiving potentiality: A metaphysics for affordances. *Topoi, 39*(5), 1177–91.

Woodward, J. (2000a). Explanation and invariance in the special sciences. *The British Journal for the Philosophy of Science, 51*(2), 197–254.

Woodward, J. (2000b). Data, phenomena, and reliability. *Philosophy of Science, 67*, S163-S179.

Woodward, J. (2002). There is no such thing as a ceteris paribus law. In *Ceterus Paribus Laws* (pp. 27–52). Springer, Dordrecht.

Woodward, J. (2005). *Making things happen: A theory of causal explanation.* Oxford University Press.

Woodward, J. (2010). Causation in biology: stability, specificity, and the choice of levels of explanation. *Biology & Philosophy, 25*(3), 287–318.

Woodward, J. (2010). Data, phenomena, signal, and noise. *Philosophy of Science, 77*(5), 792–803.

Woodward, J. F. (2011). Data and phenomena: A restatement and defense. *Synthese, 182*(1), 165–79.

Woodward, J. (2021). *Causation with a human face: Normative theory and descriptive psychology*. Oxford University Press.

Worrall, J. (1989). Structural realism: The best of both worlds?. *Dialectica, 43*(1-2), 99–124.

5
Respectful Deflationism

Edward Hall

5.1 Introduction

Gideon Rosen begins his excellent paper "Metaphysical Dependence: Grounding and Reduction" with a "plea for ideological toleration":

> Philosophers are right to be fussy about the words they use, especially in metaphysics where bad vocabulary has been a source of grief down through the ages. But they can sometimes be too fussy, dismissing as 'unintelligible' or 'obscure' certain forms of language that are perfectly meaningful by ordinary standards and which may be of some real use.
>
> So it is, I suggest, with certain idioms of metaphysical determination and dependence. We say that one class of facts *depends upon* or is *grounded in* another. We say that a thing possesses one property *in virtue of* possessing another, or that one proposition *makes* another true. These idioms are common, as we shall see, but they are not part of anyone's official vocabulary. The general tendency is to admit them for heuristic purposes, where the aim is to point the reader's nose in the direction of some philosophical thesis, but then to suppress them in favor of other, allegedly more hygienic formulations when the time comes to say *exactly* what we mean. The thought is apparently widespread that while these ubiquitous idioms are sometimes convenient, they are ultimately too 'unclear', or too 'confused', or perhaps simply too exotic to figure in our first-class philosophical vocabulary. (Rosen 2010, 109; italics in the original)

Similar pleas could be made on behalf of "metaphysically possible," "causes," "law of nature," "essence," "fundamental," "perfectly natural property," and "joint-carving," to name just a few.

Edward Hall, *Respectful Deflationism* In: *The Pragmatist Challenge: Pragmatist Metaphysics for Philosophy of Science*. Edited by: H. K. Andersen and Sandra D. Mitchell, Oxford University Press. © H. K. Andersen and Sandra D. Mitchell 2023. DOI: 10.1093/oso/9780198805458.003.0005

Surely we should respect such a plea for toleration. That said, we should accompany it with a plea for ideological *caution*. However hasty it may be to abjure some potentially useful notion out of misguided philosophical Puritanism, it is no less hasty to think that adopting that notion *requires* that we adopt some philosophically contentious gloss on it.

Example: Suppose you have become convinced that the scruples many philosophers have felt about the concept of *causation* are misguided.[1] This concept should, after all, have a place in the philosopher's toolkit: maybe it's indispensable, you think, to a good account of *reference*, or *knowledge*, or *explanation*, or *moral responsibility*, etc. Alternatively, perhaps you've become convinced that the concept of causation is somehow indispensable within scientific practice, because scientific inquiry in general aims (you think) at discovering and articulating the causal structure of the world; so a deep philosophical understanding of the nature of scientific inquiry can't help but draw on this concept. Perhaps you've found some other reason to treat causation with the utmost philosophical seriousness. One way or another, your considered view is that the word "cause" *should* "figure in our first-class philosophical vocabulary."

Fine; but you certainly haven't thereby committed yourself to, say, David Armstrong's (2001) view that the causal relation is a metaphysically primitive relation between token events. You haven't committed yourself to much of *anything* by way of a substantive conception of the nature of causation, not even that it is a mind-independent feature of the world. So far, at least, your sole commitment is highly abstract: whatever detailed account of the nature of causation turns out to be correct, it had better vindicate, in one way or another, the philosophical value you place on the concept of causation.

Simple and uncontroversial though this point is, recent work on grounding and various kindred notions has tended to obscure it from view.[2] My aim here is to offer a modest corrective, by examining, as a case study, familiar and well developed debates over laws of nature, and the best philosophical account to give of them. These debates are well developed *enough* that we will be able to see quite clearly that the decision to admit "law" into one's official philosophical toolkit leaves wide open the answer to a further, and critical, question: Do we need the word, and the concept that goes with it, in order to facilitate thought and communication about some distinctive,

[1] See, famously, Russell (1913); and for a more contemporary echo of Russell's sentiment, Norton (2003).

[2] As examples of such work, see Schaffer (2009), Rosen (2010), Fine (2012).

explanatorily important *metaphysical structure*? Or do their value for us consist in the way they play some crucial *epistemic role*, a role that can itself be characterized without reference to any such metaphysical structure? This question about the proper philosophical orientation toward a putatively essential concept generalizes, and toward the end of this essay we will briefly look at how it applies to the examples of "essence" and "ground." One crucial upshot will be that, with this kind of question in view, we can see that there was something missing from Rosen's plea. For as important as it is to be thoughtful about whether to tolerate some bit of philosophical ideology, it is at least as important to be very clear in your mind about what form this toleration shall take.

5.2 Three Stances toward a Putatively Central Concept X

To get all of this more properly in view, it will help to start with some broad methodological observations about the different stances one can take toward some putatively central philosophical concept X, and the question whether it deserves to "figure in our first-class philosophical vocabulary"; i.e., whether it has a role to play in at least some serious philosophical theorizing. Let's distinguish three stances (designed to be exhaustive);[3] we'll see that the third will need some detailed discussion, so as to forestall certain key confusions.

Robust realism: X has a role to play in at least some serious philosophical theorizing because it marks out or closely corresponds to a distinctive kind of metaphysical structure.

[3] That said, there will certainly be cases where it is hard to classify some stance. Suppose, for example, that you think that there is a metaphysically primitive kind of physical necessity that governs the world: as a matter of brute, not-further-explicable fact, some true propositions are also physically necessary, and some are not. You add that physical necessity obeys some obvious logical principles: e.g. if P entails Q, then Q is physically necessary if P is. Given your views, you're tempted to simply equate "It is a law that P" with "It is physically necessary that P." But you're aware that the way we *ordinarily use* "law of nature" stands in the way: for example, we do not conclude, from the fact that it is a law that nothing travels faster than the speed of light, that it is likewise a law that either Ned's favorite ice cream is chocolate or nothing travels faster than the speed of light. You face a choice. If you think the notion of "law" is irredeemably infected with a tendency toward such invidious distinctions among physical necessities, you should probably bill yourself as a metaphysically robust realist about *physical necessity*, but a modest eliminativist about *law*. If you think, on the other hand, that the concept of "law" can survive some modest revision, then you can go ahead and equate laws with physical necessities, and so count yourself a robust realist about both.

I am going to assume, innocently enough, that when X marks out a distinctive kind of metaphysical structure, that very fact suits it to play an important role in at least some kinds of philosophical theorizing. Less innocently, I am also going to take for granted that we have a solid enough shared understanding of what "distinctive kind of metaphysical structure" means to be going on with.[4]

Views about laws of nature provide, at least, some helpful examples; some of them will also show why that qualification "or closely correspond to" is useful. Thus, Maudlin (2007b) takes laws of nature to be metaphysically fundamental constraints on how complete physical states of the world generate successive physical states. Armstrong (1983) takes laws to be constituted by higher-order relations of necessitation instantiated by first-order universals. Laws might also correspond to a distinctive kind of metaphysical structure without themselves being fundamental, and so without the term "law" directly marking out such a structure: for example, if the entities that make up the world possess metaphysically primitive causal powers and capacities, then it might turn out that the "fundamental" laws of nature are simply regularities in the operations of these powers and capacities. That would make lawful structure metaphysically distinctive, all right—but without itself being metaphysically fundamental.

Modest eliminativism: X does not have a role to play in any serious philosophical theorizing.

A particularly illuminating example of this stance—this time, about causation—comes from Norton:

> Mature sciences, I maintain, are adequate to account for their realms without need of supplement by causal notions and principles. The latter belong to earlier efforts to understand our natural world, or to simplified reformulations of our mature theories, intended to trade precision for intelligibility. In this sense I will characterize causal notions as belonging to a kind of folk science, a crude and poorly grounded imitation of more developed sciences. (Norton 2003, 2)

Norton goes on to remind us, in effect, why the word "modest" fits: For while it is certainly *consistent* with modest eliminativism to treat a concept

[4] Though see Sider (2011) for a detailed account of what metaphysical structure itself might be.

as empty, after the manner of "witch" or "phlogiston," it's not required—and in this case, he thinks, not a good idea.

More precisely, there are many folk sciences of causation corresponding to different views of causation over time and across the present discipline. While these folk sciences are something less than our best science, I by no means intend to portray them as pure fiction. Rather I will seek to establish how their content can be licensed by our best science, without the causal notions' becoming fundamental.

For our purposes, what's crucial to modest eliminativism about X is the judgment that the concept X is not fit for serious philosophical work. In the philosophical literature on laws of nature, van Fraassen's (1989) *Laws and Symmetry* perfectly illustrates this stance.

Respectful deflationism: X has a role to play in at least some serious philosophical theorizing, but not because it marks out any distinctive kind of metaphysical structure.

It will help us going forward to pause, here, over an illuminating example. So consider one reasonable philosophical position you might take on the concept of *axiom of arithmetic*. Let's suppose that you don't for a moment think the concept should be treated dismissively, à la Norton, as "belonging to earlier efforts to understand the arithmetical realm, or to simplified reformulations of our mature arithmetical theories, intended to trade precision for intelligibility." No, the concept is much too important to mathematical practice to dismiss. And not just *mathematical* practice: the concept also, you think, needs to be drawn on when conducting serious philosophical inquiry into various topics in the *philosophy* of mathematics (e.g., concerning the nature of mathematical justification and explanation, the question what could count as a reason for *adding* to our axioms, etc.). So modest eliminativism about "axiom of arithmetic" is out.

But metaphysically robust realism about this concept should not, you think, take its place. After all, it's not remotely plausible to you that "axiom of arithmetic" applies to certain arithmetical truths, and not others, because of their distinctive *metaphysical* status. You're perfectly aware that one could try to argue that certain arithmetical truths most directly reflect or express the "essence of number," thereby securing a kind of special metaphysical status; but you think that that way of trying to deploy a metaphysically heavyweight concept of essence is a non-starter. So respectful deflationism is, you think, surely the way to go.

Good. But what precise form should it take? *Why*, if not because of its correspondence to some distinctive metaphysical structure, does "axiom of arithmetic" have a role to play in at least some serious philosophical theorizing? Let's suppose that at this point you turn to certain *epistemic* features of the concept. You point out that certain arithmetical truths strike us as especially obvious or "self-evident." In addition, certain truths are, at least in conjunction with one another, logically fecund, in that it is possible to deduce a large and heterogeneous set of other truths from them. And some such logically fecund sets are pleasingly minimalist, in that each truth contained in them is essential to their fecundity. You propose, then, that what makes some arithmetical truths deserve to count as "axioms" is precisely that they each strike us humans as at least as immediately obvious as any other arithmetical truth, and that they are collectively fecund in a minimalist fashion. That would be a way to make good on the promissory note built into the definition of "respectful deflationism": namely, that as a respectful deflationist about X, you have some *other* way of explaining why we should include X in our "first-class philosophical vocabulary." Note, finally, that in this case there is a clear sense in which your way of explaining is *pragmatist* in that the value you place on the concept "axiom of arithmetic" does not hinge on that concept's correspondence to some distinctive metaphysical structure, but rather on the way that it helps to structure mathematical practice so as to serve certain basic interests we share, in engaging in that practice.

It is certainly not built into the definition of respectful deflationism that it will take this pragmatist orientation. But I think it very often will—and, indeed, that if you find yourself wanting to include some concept X in your first class philosophical vocabulary, but at the same time find robust realism about X unattractive, then the best strategy for you is to look for some deep and pervasive interest that would be served by having a concept that works in the way that X appears to.

5.3 Two Mistakes about Respectful Deflationism

You don't have to think that the stance toward the concept of an axiom of arithmetic sketched above is correct; it's enough for our purposes to see it as coherent and not obviously mad.[5] It will serve us not just as one illustration

[5] Though even if it's on the right track, it almost certainly needs to be augmented, by considering much more systematically the different kinds of interests we have in engaging in

of a plausible respectful deflationism, but as an occasion for singling out and diagnosing two serious potential confusions about it and many other instances of respectful deflationism, confusions that can easily arise when that stance toward some concept X takes a broadly pragmatist form. (Later, we'll see both confusions at work in the literature on Humeanism about laws of nature.)

To see the first confusion, go back to this claim from the last section (with some added emphasis):

> what *makes* some arithmetical truths deserve to count as "axioms" is precisely that they each strike us humans as at least as immediately obvious as any other arithmetical truth, and that they are collectively fecund in a minimalist fashion.

It can look, superficially, as though what's being signaled here by the use of "makes" is that this is a *grounding* claim, a *metaphysical explanation* of what it is for some arithmetical truth to be an axiom. That's confused; but it's a confusion smoothly engendered by the kind of language we typically use in explications of this sort. Thus, we could just as readily have said that the fact that a certain truth is an axiom *depends upon*, or *is grounded in*, or holds *in virtue of*, or is *made true by*, the fact that that truth belongs to a set of arithmetical truths meeting the foregoing conditions. All such idioms—which, note, are the very idioms on whose behalf Rosen issued his plea—naturally suggest that we are saying what *metaphysically explains* a given truth's status as an axiom.

Again: this is a confusion. Or perhaps more carefully: there is no compelling reason to think that the view about "axiom of arithmetic" that we have been sketching needs to market itself as offering a metaphysical explanation of axiom-hood. To see how it *should* market itself, we need to shift gears, and consider a different question. Suppose that our starting point is that as far as metaphysical status is concerned, all arithmetical truths are exactly on a par. Of course, there will be local logical asymmetries among them: P may

mathematical inquiry. For example, suppose that your proposed set of axioms features the claims P and Q, and my proposed set features the claim R—where R is provably equivalent to the conjunction of P and Q. Suppose in addition that R is attractively simple and elegant; i.e., it's not just (P & Q). Suppose our axioms are otherwise identical. Still, we might end up preferring *your* axioms, precisely because they allow us to distinguish results whose proofs depend on P but not Q from results whose proofs depend on Q but not P; and we might find such distinctions mathematically *illuminating*. If so, that would show that our desiderata of obviousness and fecundity are incomplete, in ways that will matter quite a lot to how we structure our practice. And of course there may be other desiderata as well.

be logically deducible from Q but not conversely, etc.[6] But there is no interesting metaphysical sense in which certain truths of arithmetic are "more fundamental" than others. We now face an interesting question: What is the *rationale* for having a concept of "axiom" that functions in the way ours seems to (namely, by singling out certain arithmetical claims as playing a special role in arithmetical justification)? Or to put the question in a more usefully evocative form: If we didn't already *have* the concept of "axiom," what reasons would our arithmetical practices give us for *inventing* it?

And to that question, we can give a compelling answer. We have, after all, various *interests* when it comes to arithmetic. Even setting aside practical applications, we have an interest in extending our arithmetical knowledge, and, perhaps, in extending our arithmetical understanding. (I have in mind here the thought that two proofs of the very same result can differ with respect to how *illuminating* they are, and *more* illuminating proofs strike us as more valuable. That fact seems to indicate that more than just an interest in increasing our store of established results motivates mathematical inquiry.) Fortunately, we don't start out in a hopelessly impoverished state: certain arithmetical claims strike us as blindingly obvious—too obvious to warrant any serious skepticism about whether we are justified in believing them—and we have available to us principles of logic that we trust to transmit this kind of warranted certainty. We could, of course, proceed in a piecemeal and individualistic fashion: where you yourself find some claims to be obvious, set about trying to see what you can deduce from them, never mind what the rest of us are doing. But that is a disorganized approach inimical to the pursuit of arithmetical knowledge and understanding as a collective enterprise. So it makes good sense for us to single out certain arithmetical claims to play a privileged role in our joint arithmetical practice (e.g., via a prevailing norm that any mathematical justification be, at least in principle, capable of being recast as a deduction from *those* claims). The point of *having* a concept of "axiom" is simply to have a way to think and talk about *those claims*. And which claims those should *be*—which claims, that is, end up being designated "axioms"—should be settled pragmatically, by reference to the *point* of structuring our collective practice of arithmetical inquiry around them in a certain way.

[6] Note that we're assuming, in the background, that the arithmetical truths are either expressed in some canonical language for which a notion of logical deducibility is well defined, of that they somehow have an intrinsic syntactic structure that permits a notion of deducibility to be defined for them in a language-independent fashion.

With this background in place, we can give a clean diagnosis of the confusion indicated above. It is perfectly legitimate to ask, of those claims that have been designated "axioms," this question: why are those claims axioms? It is legitimate, moreover, to rephrase in any number of ways: In virtue of what are those claims axioms? What makes it the case that those are the axioms? And so on. But whatever the phrasing, this question admits of two very different interpretations. One interpretation treats it as a request for a metaphysical explanation. That's confused. The *right* interpretation treats it as a request for the *rationale for a decision*. That is quite a good question, albeit one better, because much less misleadingly, phrased thus: given our interests in arithmetical inquiry, what purpose is served by singling out certain arithmetical claims for special treatment, in the way the word "axiom" does?

It is a serious error to confuse a rationale for a decision for a metaphysical explanation. I think it's an error comparatively easy to spot, in the case we've just been discussing. By contrast, in the literature on Humean approaches to laws of nature, the error is widespread, and, for reasons we will come to, more difficult to spot. But we will also see that once it is undone, the way is open to appreciating just how radically pragmatist, anti-metaphysical, and well-motivated the kind of respectful deflationism exhibited by Humeanism about laws of nature can be. And that insight will lay the groundwork for future work, aimed at showing how respectful deflationism of this variety can extend to "ground," "essence," and perhaps many other metaphysical categories.

The second confusion to which respectful deflationism is vulnerable, especially when it takes a pragmatist form, is that it is somehow unacceptably "subjective" or "anthropocentric." Interestingly, Lewis makes exactly this kind of complaint about his own "best system" account of laws of nature, which (as we'll see) is a paradigm of respectful deflationism. The complaint arises, he thinks, because of the way his account privileges the notions of simplicity, informational strength, and ways of trading them off against each other:

> The worst problem about the best-system analysis is that when we ask where the standards of simplicity and strength and balance come from, the answer may seem to be that they come from us. Now, some ratbag idealist might say that if we don't like the misfortunes that the laws of nature visit upon us, we can change the laws—in fact, we can make them always have been different—just by changing the way we think! (Talk

> about the power of positive thinking.) It would be very bad if my analysis endorsed such lunacy. (Lewis 1994, 479)

Unfortunately, the clever Lewisian rhetoric makes it hard to see what exactly the problem is supposed to be. I think we can bring it into sharper focus—and see, in fact, how insignificant it is—by returning to our working example of axioms of arithmetic.

The alleged problem about subjectivity can take at least three forms. First, one might worry about *variation between people*. Thus, what strikes one person as so obvious as to stand in no need of justification might strike another as demanding proof. More subtly, judgments of logical fecundity might vary between people: for these judgments can't be based on the mere cardinality of the logical consequences of a given set of arithmetical claims, since the cardinality will be the same (countably infinite) in each case. Plausibly, they rest instead on how "diverse" and "interesting" are the results that can be deduced from some candidate set of axioms, and judgments about diversity and interestingness might themselves vary across people. And there may be other desiderata on axioms, judgments about which can vary (cf. footnote 5, above). Conclusion: the account on offer yields the unacceptable conclusion that people could *faultlessly disagree* about what are the axioms—faultlessly, because the account gives us no basis for saying that where there is such disagreement, at most one of the disagreeing parties is correct.

Second, one might worry about *undue control*. Lewis seems to have this worry in mind, when he writes about "changing the laws." In our working example, the worry might be that we can control which truths are the axioms of arithmetic, simply by deciding to pursue different aims in arithmetical inquiry. But surely it is not just "up to us" in this way what the axioms of arithmetic are.

Finally, one might worry about *objectivity*. Lewis also seems to have this worry in mind, as witness his own response to his challenge:

> if nature is kind to us, the problem needn't arise. I suppose our standards of simplicity and strength and balance are only partly a matter of psychology. It's not because of how we happen to think that a linear function is simpler than a quartic or a step function; it's not because of how we happen to think that a shorter alternation of prenex quantifiers is simpler than a longer one; and so on. Maybe some of the exchange rates between aspects of simplicity, etc., are a psychological matter, but not just anything

> goes. If nature is kind, the best system will be *robustly* best—so far ahead of its rivals that it will come out first under any standards of simplicity and strength and balance. We have no guarantee that nature is kind in this way, but no evidence that it isn't. It's a reasonable hope. Perhaps we presuppose it in our thinking about law. (Lewis 1994, 479)

Unfortunately, Lewis doesn't tell us what he thinks simplicity *is* a matter of, if not psychology. But presumably he thinks that the one true theory of simplicity (and strength? and balance?) had better explain it at least partly in terms of factors that have nothing to do with psychology, or else his account of laws will, unacceptably, yield the consequence that it is not an objective fact what the laws are. Similarly, our working account of axiomhood may seem to yield the unacceptable consequence that it is not an objective fact what the axioms are.

To see how best to respond to these concerns, let's consider what, in general, the *point* is of having concepts at all. It's an absurdly broad question, of course; and it's just as clear that there is more than one answer. For our purposes, it will be enough to distinguish two answers, each appropriate in a different range of cases. The first answer begins from the somewhat inchoate but highly suggestive idea that one of our aspirations, in conducting inquiry into some subject matter, is that our thoughts *fit* that subject matter in the right way. I was implicitly appealing to this idea when I assumed, above, "that when X marks out a distinctive kind of metaphysical structure, that very fact suits it to play an important role in at least some kinds of philosophical theorizing." Sider gives an evocative statement of this view:

> The goal of inquiry is not merely to believe truly (or to know). Achieving the goal of inquiry requires that one's belief state reflect the world, which in addition to lack of error requires one to think of the world *in its terms*, to carve the world at its joints. Wielders of non-joint-carving concepts are worse inquirers. (2011, 61; italics in the original)

Given some concept X, then, *one* point to having this concept is that it facilitates this goal of inquiry: for only by thinking by means of it are we able to think about the given subject matter "in its terms." Obviously, this answer is highly congenial to robust realism about X (and will entail it outright, if Siderean "joints" just are distinctive kinds of metaphysical structure).

Still, a given inquiry might have *many* goals, not exhausted by Sider's twin aims of believing truly and in a joint-carving manner. These goals might have any number of sources, and depending on these sources prove more or less malleable within individual people, and more or less variable across people. Once they are in place, though, a very different answer to our "what is the point?" question becomes available: it might be that the point of having a given concept X is that the content and method of deployment of this concept facilitates our goals in some significant way.

Suppose that, in the case of some given concept X, this answer is apt. Respectful deflationism about X does not follow. That's because the details matter. Remember that respectful deflationism is *respectful*: X needs to have some role to play in serious philosophical theorizing. There are all sorts of reasons why, even though possessing X helps us attain some important goal, X might fail to meet this further standard. Giving an exact and complete theory of these reasons looks to me to be quite difficult; but we can at least point to some important considerations. For example, it might be that the goal in question is too malleable: we could easily change it in such a way as to render X pointless. Or it might be that there is too much variation across people in the way in which possession of X facilitates the goals of the given inquiry. At any rate, I think we have said enough to see how to respond to the worries raised above about subjectivity.

Let's take them in reverse order. The third complaint was that our proposed treatment of axiomhood has the consequence that it is not objective what are the axioms of arithmetic. The complaint is question begging. To take it seriously, we would have to have been antecedently convinced that the point of having a concept of "axiom" is that it helps us think about the numbers "in their own terms," by carving numerical reality at one of its joints. Respectful deflationism about axiomhood rejects this picture, right at the outset. So of course it is going to reject the view that what the axioms are is an objective matter.

And this failure of objectivity does not look so troubling, once we answer the other two concerns. To begin, the worry that it might be "up to us" what the axioms are seems quite misplaced. The basic interests that motivate inquiry into arithmetical truth may not be widely shared among humans (i.e., most humans can't be bothered to study number theory), but where they *are* shared their specific character does not seem malleable at all. Nor—to respond to the first concern—does the character of this interest seem highly variable across people. Anyone who takes an interest in number theory at all is going to find, for example, Goldbach's conjecture interesting; and no one is going to find similarly interesting the question, say, of how many times the number 7 appears in the digital decimal expansion of 12,345^67,890.

Of course, we can and should hold it out as a remote empirical possibility that there are people—perhaps many of them—whose judgments about which arithmetical claims are most interesting to investigate is very different, and perhaps even people whose judgments about which arithmetical claims are so obvious as to stand in no need of justification are likewise very different. But the really important point is that even if this unlikely state of affairs turned out to obtain, it would do nothing to show that *robust realism* about axiomhood was correct; what it would show, instead, would be either that the concept of axiom does not, after all, have an important role to play in structuring mathematical inquiry, or that the way it does so is highly variable and inconstant. Either way, we would need to conclude that the "respectful" part of our respectful deflationism was misplaced: given how very differently it turns out different humans think about arithmetic, the concept of "axiom of arithmetic" is not, after all, suited to be part of our first-class philosophical vocabulary. But that is as it should be. For respectful deflationism about *any* concept is going to be, in this way, very modestly hostage to empirical fortune. And should these fortunes go awry, it is modest eliminativism that we must fall back to, not robust realism.

There is a further point worth emphasizing, about the charge of subjectivity and the dialectical situation between the robust realist and the respectful deflationist. In many cases—though not invariably—robust realism about X and respectful deflationism about X will be quite intimately connected, in a specific way. Namely: there will be certain considerations that the robust realist about X will take as our *evidence* that the concept X applies, in a given case; and the respectful deflationist will draw on those *very same considerations* in giving her rationale for constructing X in the way she does. When the two stances are connected in this way, the charge of subjectivity not only fails to advance the robust realist's case, but seriously undermines it.

That's all quite abstract, so let's see the point in action. Billy is a robust realist about "axiom of arithmetic," of the kind we saw above: he thinks that a certain elite set of arithmetical truths directly express the metaphysical essence of number, and that *that* is what earns them their status as axioms. Suzy subscribes to the view we've been sketching, according to which nothing metaphysical distinguishes axioms; rather, they earn their place because of the way that structuring arithmetical inquiry in terms of them serves our aims in conducting that inquiry. For reasons we've seen, Suzy places a great deal of emphasis on judgments of obviousness, and on judgments of mathematical fecundity. *And so does Billy*—but in *his* case, these judgments are what we must rely on in order to *discover* which of the arithmetical truths are the axioms. So the two views are at once intimately connected, and profoundly different in orientation: for Suzy, the *reason it's a good idea* to denominate certain arithmetical claims "axioms" is that they collectively exhibit the right kind of obviousness and fecundity; for Billy, the *reason we should consider it likely* that they are the axioms is that they collectively exhibit the right kind of obviousness and fecundity.

Now suppose Billy complains, à la Lewis, that Suzy's view is open to the problem of "ratbag idealism": we could go ahead and arbitrarily change the way we judge obviousness and fecundity, thereby changing the facts about what the axioms are. Or maybe he's not worried about the malleability of these judgments, but about the way they depend on idiosyncratic features of human psychology. Either way: lunacy!

Now, we've already seen why Suzy should be unmoved. We humans likely have much less room for maneuver with respect to these kinds of judgments than the challenge seems to imply; and at any rate, the worst that can happen is that we discover that the significance of the concept of "axiom" is more local and contingent than we might have thought. As for the charge of idiosyncracy, she should embrace it: why is it supposed to be a problem for her, if it turns out that the value of the concept of "axiom" consists in the way it usefully structures distinctively *human* arithmetical inquiry?

But Billy should be *very* concerned about the significance of the worries he's raised, for his *own* position. Consider just the charge of idiosyncrasy. The worry is supposed to be that our judgments concerning which arithmetical claims are "obvious" and which are "interesting" (in the sense relevant to settling the fecundity of axioms that can decide them) reflect peculiarities of the human cognitive system. But if so, why think that these judgments serve as reliable guides to the special metaphysical status that the axioms are supposed to enjoy? By opting for robust realism about axiomhood, Billy already added significantly to the burdens placed on his epistemology. Now it looks as though he's made them insurmountable.

Let's turn now to a close look at an additional and highly illuminating case study: contemporary Humean treatments of the concept "law of nature."

5.4 Case Study: Humean Approaches to Laws of Nature

Contemporary "Humean" accounts of laws of nature all attempt, in one way or another, to make good on the guiding idea that laws of nature are nothing more than certain kinds of distinctive patterns in the phenomena. I am going to skate over the fascinating history that led to the current most sophisticated forms of Humeanism about laws, and instead present a general framework that I think does a good job of capturing their common features. Here, then, is a recipe for constructing such an account.

We begin with a set of canonical languages, sentences of which can be used to express propositions about the world. Let's take immediate note of two points. First, we use the plural "languages" precisely because we want to allow a Humean the option of constructing her account of laws in an explicitly pluralistic fashion; more on this in a moment. Second, while Humeans should certainly engage in healthy debate amongst themselves about what a language has to look like in order to count as "canonical," all should agree on one constraint: the basic vocabulary of such a language should be entirely non-modal, in that, roughly, it cannot be used to directly express any concepts of physical modalities such as law of nature, cause, chance, etc. (In a moment, we'll see a better way to capture this constraint.)

Given this starting point, our Humean will now consider, for each language, all the possible sets of sentences in that language. As her second crucial ingredient, she will then impose *constraints* on these sets. As before, there should be healthy debate amongst Humeans about what these constraints should be; but all should agree that they must be specified in a way that makes no reference to physical modalities. For example, it will not do to say that a candidate set "passes" only if all of the sentences in it express truths about the world that hold as a matter of physical necessity.

Finally, our Humean will need to make some decision about what status to assign to those sets that *meet* all of her constraints. If there is just one such set, the decision is easy: the sentences in this set are all and only the laws of nature. If there is just one such set for each of the languages of her canonical languages, pluralism recommends itself: e.g., if one language is particularly apt for expressing chemical facts while another is particularly apt for expressing physical facts, then the "best system" of sentences of the first language will contain the chemical laws, while the best system for the second language will contain the physical laws. If multiple sets within a given language are tied for "best," or if there is an infinite sequence of successively "better" sets, she will have more work to do—perhaps she will want to say that it is indeterminate as between these sets what the laws are, or that the laws are given by all and only those sentences common to all winning sets, or something else.

Let's see how all of this plays out, in Lewis's own presentation of his "best system" account.[7] First, Lewis takes there to be just *one* canonical language: it is a first-order language whose non-logical vocabulary expresses the "perfectly natural" properties and relations instantiated in our world. I won't go into Lewis's account of what perfectly natural properties are (though see his 1983), save to note two key points: first, according to Lewis, no two metaphysically possible worlds differ with respect to what is qualitatively the case without differing, somehow, in the pattern of instantiation of perfectly natural properties and relations at them. Second, these properties and relations qualify as "non-modal" because they obey a principle of recombination: roughly, any combinatorially possible pattern of instantiation of them is metaphysically possible. (This is the better account of the basic constraint on canonical languages alluded to above.) Lewis then stipulates that a "best" set of sentences in this language must include only sentences that are true, and must in addition achieve an optimal balance of simplicity and informativeness. Unfortunately, he says very little about how simplicity, informativeness, and the trade-offs between them should be understood; but some of his writings suggest that he thinks that we can at least point to these standards, as the standards that scientists in fact use in deciding the relative credibility of rival theories.

Much work since Lewis has increased our understanding of the range of alternative Humean views that are available, and that is precisely why I laid out the recipe in so generic a fashion. For example, Cohen and Callender (2009) explicitly take up the pluralist option I indicated above. My own (2007) worries that informativeness can't be an appropriate standard to impose on candidate systems, and so worries that the best system account itself is untenable; Hicks (2018) provides a decisive response, by laying out a very different and quite plausible set of constraints on candidate systems that neatly evades my worry. Albert (2012) contains an illuminating discussion of how concepts from statistical mechanics can be used to articulate important kinds of constraints on candidate systems. Loewer (2007) explores ways in which canonical languages can be

[7] See for example Lewis (1983).

constrained that departs radically from Lewis's reliance on the metaphysically heavyweight distinction between natural and non-natural properties and relations. And this is just a quick sampling: the last two decades, especially, have seen a profusion of highly creative philosophical work on Humean accounts of laws of nature, all of which conforms to the recipe sketched above.

But I think that this work has also entrenched a serious philosophical mistake, which is that what a Humean account of laws is supposed to be offering is, among other things, a *metaphysical explanation* of facts about laws in terms of other, suitably non-modal facts. Consider this claim from a recent paper by Loewer (a card-carrying Humean about laws); he is offering it as a rebuttal to an argument advanced in Maudlin (2007b) and elsewhere, that the Humean's laws can't explain particular non-modal facts because those laws *are themselves explained by* such facts:

> I claim that this objection rests on failing to distinguish metaphysical explanation from scientific explanation. On Lewis' account the Humean mosaic metaphysically determines the L-laws [L-laws = laws, as a Humean understands them]. It metaphysically explains (or is part of the explanation together with the characterization of a Best Theory) why specific propositions are laws. This metaphysical explanation doesn't preclude L-laws playing the usual role of laws in scientific explanations.
>
> (Loewer 2012, 131)

Lange (2013) resuscitates the charge of explanatory circularity, by arguing that the two kinds of explanation—metaphysical and scientific—should obey a transitivity principle, whence Loewer's maneuver won't help to evade the charge; Lange simply assumes without comment that the Humean account is *meant* as a metaphysically explanatory grounding thesis. This same assumption appears to hold in most of the published responses to Lange (Miller 2015 being a notable exception).

As noted, I think this is a serious mistake. Or—to be a little more concessive—that there is absolutely no reason why a Humean needs to present her account as such a grounding thesis. Hopefully, the alternative is by now obvious: she can present her account as, first and foremost, an account of the *rationale* for having a concept of law that works in a certain way to organize our empirical inquiry.

In fact, we can spot a couple instances in the literature where the author tiptoed up to precisely this understanding of what the Humean should take herself to be up to. For example, in Beebee's excellent paper "The Non-Governing Conception of Laws of Nature," she writes:

> So the idea is something like this. Suppose God wanted us to learn all the facts there are to be learned. (The Ramsey-Lewis view is not an epistemological thesis but I'm putting it this way for the sake of the story.) He decides to give us a book – God's Big Book of Facts – so that we might come to learn its contents and thereby learn every particular matter of fact there is. As a first draft, God just lists all the particular matters of fact there are. But the first draft turns out to be an impossibly long and unwieldy manuscript, and very hard to make any sense of – it's just a long list of everything that's ever happened and will ever happen. We couldn't even come close to learning a big list of independent facts like that. Luckily, however (or so we hope), God has a way of making the list rather more comprehensible to our feeble, finite minds: he can axiomatize the list. That is, he can write down some universal generalizations with the help of which we can derive some elements of the list from others. *This will have the benefit of making God's Big Book of Facts a good deal shorter and also a good deal easier to get our rather limited brains around.*
>
> (Beebee 2000, 574; italics added)

A rationale-focused Humean knows exactly how to read this passage: the final sentence provides a reason why, given our limited cognitive abilities together with our interest in knowing as much about the world as possible, it makes sense for us to single out certain truths for special attention, and so denominate them "laws." This focus on rationale is even more vivid in Albert's "Physics and Chance":

> Here's the idea. You get to have an audience with God. And God promises to tell you whatever you'd like to know. And you ask Him to tell you about the world. And He begins to recite the facts: such-and-such a property (the presence of a particle, say, or some particular value of some particular field) is instantiated at such-and-such a spatial location at such-and-such a time, and such-and-such *another* property is instantiated at such-and-such *another* spatial location at such-and-such another time, and so on. And it begins to look as if all this is likely to drag on for a while. And you explain to God that you're actually a bit pressed for time, that this is not all you have to do today, that you are not going to be in a position to hear out the whole story. And you ask if maybe there's something meaty and pithy

> and helpful and informative and short that He might be able to tell you about the world which (you understand) would not amount to everything, or nearly everything, but would nonetheless still somehow amount to a lot. *Something that will serve you well, or reasonably well, or as well as possible, in making your way about in the world.*
>
> (Albert 2012, 23; last italics added)

The overarching idea is straightforward, and is straightforwardly an *alternative* to the "grounding" approach. We humans have some basic interests that both motivate and structure the way we conduct empirical inquiry. Given her distinctively deflationary metaphysical background, the Humean is already committed to thinking that, insofar as these basic interests should be vindicated rather than set aside as confused, they can be accurately characterized in non-modal terms.[8] It may well turn out (although, to be sure, this is hardly obvious ahead of time) that certain claims about the world have an especially high value, from the standpoint of these aims: for short, they are *distinctively appropriate targets for scientific inquiry*. Of course, a substantial amount of philosophical work needs to be done both to carefully articulate the aims, and to argue that certain sorts of claims about the world will, if true, *have* this distinctive value. That's what makes the project of constructing a Humean account of law interesting, and difficult. But once all this is accomplished, the final step is simplicity itself: the point of having a concept of "law" is simply to have an efficient way to think and talk about *those claims*. Of course the Humean is explaining *something*: she is explaining why, given our interests, it is important for us to single out for attention certain features of reality. But it is just a confusion to think that that is any kind of *metaphysical* explanation of *anything*.

I think there are at least two reasons why this sensible way of characterizing the Humean project has come to be overlooked. The first is the sloppiness of our language noted above: we can readily if loosely express what the Humean is up to by saying that she is giving an account of what "makes" something a law, or that she is showing us that certain claims are laws "in virtue of" having such and such features, and so on. All the points made

[8] The qualification is necessary. After all, some scientists have explicitly stated their interests in ways that the Humean cannot respect—e.g. as an interest in discovering the fundamental rules that God has laid down for how the physical world shall behave. So a Humean is of course going to have to defend her view in part by rejecting such motivations as confused, and as needing to play no role in a careful rational reconstruction of the legitimate aims of scientific inquiry (Hall 2007).

above about "axiom of arithmetic" apply here, so I won't belabor them; suffice it to say that the crucial terminology does not carry its interpretation on its face, and one point of distinguishing a rationale from a metaphysical explanation is to expose just this unclarity.

I think there may be a second reason, as well. Lewis originally stated his best system account not as an account of the rationale for having a concept of law that worked in such and such a way, but rather as a supervenience thesis. So right at the outset, he set up the dialectic in a way that obscured the former option from view. And in the subsequent years, metaphysics in general began to turn away from supervenience as an appropriate tool for capturing the structure of certain kinds of accounts, so that by now it has become commonplace to observe that supervenience is no substitute for metaphysical grounding. So the thought in the back of people's minds may have been that Humeanism about laws is, first and foremost, supposed to be a certain kind of thesis about the *metaphysical status* of laws, and that, since this metaphysical status obviously can't be adequately captured by means of supervenience, it needs instead to be captured by means of the notion of metaphysical grounding.

But if I am right, the mistake occurred right at the beginning. It is not that the Humean doesn't operate from a distinctive metaphysical perspective. Of course she does: hence, among other things, the non-negotiable constraint on which languages can count as "canonical," and the non-negotiable constraint on how one is allowed to articulate the constraints on "best" systems. But from that point on, her metaphysical principles have done all the work they need to do. She uses those principles to identify a certain range of facts about reality as a suitable totality of non-modal facts. And then she asks, "What rationale could there be for distinguishing certain of these facts as *special* from the standpoint of scientific inquiry, in the way that our use of the term 'law' seems to be trying to pick out?" Her metaphysics enters here, as well—but only in her insistence that the distinctive aims that a concept of law enables us to pursue can themselves be characterized without reference to the kind of modal structure she rejects. On this approach, supervenience is an *afterthought*. Put another way, it's not as though the Humean thinks to herself, "Well of course there are modal facts—facts about laws, causes, chances, etc. So my job, as a Humean, is to figure out how those facts supervene on, or are metaphysically grounded in, some appropriate range of kosher, non-modal facts." It is rather that there are *just* the non-modal facts, together with distinctive interests we have as inquirers that provide us with good reason to single out some of them for special attention.

Taking this rationale-focused approach to Humeanism confirms some clear benefits. Here are three that I think especially worth mentioning. First, the explanatory circularity challenge—which has attracted quite a lot of attention in the literature, in recent years[9]—dissolves immediately. Not being *metaphysically* explained by anything, laws certainly are not metaphysically explained by any fact that, in turn, we might hope to scientifically explain by means of them.

Second, Lewis's concerns about the "ratbag idealist"—which, remember, he himself labeled as the most serious problem for his account of laws—are revealed as wholly baseless. Here we can simply rehearse analogs to the points made about "axiom of arithmetic," in the last section. To begin, the charge that the standards for which candidate systems count as "best" may turn out to depend on idiosyncratic features of human psychology, and so yield the result that it is not an objective fact what the laws are, is question begging: as soon as we take the stance that the critical question to ask is what *point* there is for creatures like us, with distinctive epistemic interests that motivate and structure inquiry into the world around us, to make use of the concept of law, we have already set aside a conception of "law" as limning some sort of objective metaphysical structure. And it would be daft for us nevertheless to insist that whatever rationale we can point to for introducing the law/non-law distinction cannot make any reference to central features of the human condition that bear on what sorts of basic epistemic interest we have. Next, should it turn out that the basic interests that motivate the need for a concept of law vary too much between people, or across time, then the proper response is to become modestly deflationist about the concept. Robust realism is no refuge. If anything, the situation will be much more awkward for the robust realist. After all, the kind of variation in basic epistemic interests we are contemplating will lead the respective communities to make profoundly different but equally clearheaded judgments about which sorts of claims about the world deserve to be called "laws." The robust realist will have to maintain, somehow, that at most one of those communities can be correct. Lucky them!

Finally, I think that a focus on what could be a Humean-friendly rationale for having a concept of law is philosophically very healthy, in that it directs attention to questions of central philosophical importance that could otherwise be too easily overlooked. It is an oddity of Lewis's own work, for

[9] See for example Armstrong (1983), Maudlin (2007a), Loewer(2012), Lange (2013), Miller (2015), Marshall (2015), Hicks and van Elswyk (2015), Lange (2018).

example, that he said so very little about what simplicity and informativeness are, or how a best system should balance them. A focus on rationale leaves no room for such quietism. How can you pretend to be explaining the point of having a concept of law that functions the way Lewis's does, if you say nothing about nothing of substance about how the pursuit of simplicity and informativeness meets our needs as epistemic inquirers?

I've argued that Humeanism about laws is safe from two challenges—the ratbag idealist and explanatory circularity—widely thought to pose serious threats. Real challenges remain. The best place to look for them is in potential mismatches between what the *Humean* must put forth as the most coherent rationale for a concept of law, and what should *actually* be put forth. Suppose, for example (and very plausibly), that one of our core interests as inquirers is not merely to gain more knowledge about what the world is like, but to gain *explanations*. And suppose further (and now much more controversially) that the only way to make sense out of this interest in explanation is to see it as an interest in knowledge of *objective constraints* on the phenomenon or phenomena to be explained. So in particular, our interest in laws as components of explanations only makes sense, on this view, if laws act as objective constraints on how the world can be. This very understanding of what we are after in inquiry is one the Humean cannot accept. So if it is correct, then of course the Humean account of laws is bound to fail (unless, somehow, laws are wholly irrelevant to explanation). The lesson here is that a shift to a focus on the rationale for the concept of law does not necessarily make it easier to defend Humeanism. What it *does* do is to direct our attention to where the genuine challenges will lie.

5.5 Speculations about "Essence" and "Ground"

Let me turn now to some brief remarks about how the discussion so far might point the way toward a kind of respectful deflationism toward essence and ground.

Begin with essence. We can inquire into the essence of a particular entity; we can also inquire into the essence of a kind. In each case, one route into respectful deflationism about essence begins by reinterpreting a request for the essence of a thing or a kind as a request for information that will be distinctively valuable in organizing subsequent inquiry focused on that thing, or that kind.

Suppose, for example, that you are interested in black holes. But it's not just that you want any old information about black holes—and so, for example, would be satisfied to hear that there is probably one at the center of our galaxy, or that Stephen Hawking has written about them, or that they feature prominently in many science fiction movies, and so on. No; what you want to know is, as you might put it, what is *essential* to black holes *as such*. A metaphysically robust realist about essences of kinds will take this question at face value: you just told us, straightforwardly and in appropriately metaphysical terminology, what sort of information you wanted. A respectful deflationist, by contrast, will take a slightly more subversive approach, executing a Euthyphro-like maneuver here. What she will say is that, insofar as you have an interest in conducting further inquiry about black holes—being able to answer questions about them, being able to absorb new information about them in an effective and organized manner, etc.—then there will be certain facts about black holes that will be of distinctive value to you, and certain other facts that will be vastly less important. For example, it will be much more value valuable to you to know about event horizons then it will be for you to know about the number of black holes within 100 light years of earth. Crucially, the respectful deflationist needs to think that the value this information has for you can be characterized *without invoking a metaphysically heavyweight notion of essence*. Here, in fact, is precisely where the Euthyphro maneuver comes in: according to her, it's not that certain information is valuable for inquiry because it is information about what is essential to being a black hole, it is rather that we label certain information about black holes that way only because it has a distinctive sort of value for black-hole-directed inquiry. In short: our interests as inquirers about black holes provide a rationale—a rationale that can be characterized without invoking any sort of distinctive metaphysical structure—for drawing a distinction between certain facts about black holes and certain other facts. The word "essence" merely serves as a way to mark this distinction.

Consider now the concept of ground. Here it is much more difficult to sketch a simple respectful deflationist approach, but that is mainly because the concept itself is deployed in such a heterogeneous fashion. Thus, we can find philosophical speculations about whether macroscopic properties of physical objects are grounded in their microphysical constitutions; about whether the normative is grounded in the non-normative; about whether structured facts are grounded in their atomic constituents; and so on. I very much doubt that there will be a respectfully deflationist approach to "ground" that uniformly covers all such examples. But as one illustrative example, consider the case of structured facts. These are often held up as paradigmatic examples of where a *metaphysical* relation of grounding should be taken to obtain. Here for example is Rosen:

> Some of the clearest examples of grounding involve facts that stand in simple logical relations. Thus it seems quite clear that if there are disjunctive facts, then a disjunctive fact is grounded in its true disjuncts. If Fred is in New York, then Fred is either in New York or Rome. Moreover, the fact that Fred is in New York or Rome obtains *in virtue of the fact that Fred is in New York*. (Rosen 2010, 117)

But I think it is possible to take a very different approach even to this allegedly compelling example, though it is easiest to see this possibility if we shift focus from structured facts, for which there is no notion of falsity, to structured sentences, which can be either true or false. Suppose then that we have some simple language, consisting of atomic sentences and all sentences that can be formed from these by means of the usual sentential connectives. Let this be an interpreted language, so that each sentence in it is either true or false. It's very tempting to endorse Rosen-like intuitions, and insist that, for example, the truth or falsity of

$$(p \vee q) \;\&\; r$$

Is immediately grounded in that truth or falsity of the pair

$$(p \vee q),\, r$$

And that the truth or falsity of these sentences is grounded in the truth or falsity of the atomic constituents p, q, and r.

But there is another way to go, reminiscent of the respectful deflationist account we sketched of the concept of axiomaticity. There are, to begin, relations of logical implication between the sentences in this language. And thanks to these relations of logical implication, there will be additional relations that we might capture by the following schema:

"That—is true/false guarantees that—is true/false."

where the first blank can be filled in by multiple sentences, and the second by a single sentence. But this relationship exhibits no striking asymmetries, so it won't do as a candidate for "ground." For example, suppose it is the case that the atomic sentence p is true. Then these are also the case:

That p is true guarantees that ¬p is false.

That ¬p is false guarantees that p is true.

Now, suppose we are interested in organizing our view of the truth values of the sentences of this language in as cognitively efficient a manner as possible. Then one thing that might be valuable to us is a specification of a subset of the sentences such that their truth values determine the truth values for all other sentences in the language, no matter how all those truth values are in fact distributed. We might, further, find it attractive for this set to be minimal: no proper subset of it has this guaranteeing feature. There will be many ways to satisfy these constraints. For example, if our language contains just the atomic sentences p, q, r, and s, then this set will do:

(p iff q), (p iff q iff r), (p iff q iff r iff s), (s iff p)

But if, in addition, we want a set that *we*—given our very human cognitive limitations—can use as *effectively and efficiently* as possible, then this last choice is obviously a poor one. What's the best choice? You already know: it's just the set of atomic sentences. And we might mark off the special status this set has by saying that it is the truth values for sentences in it that are the "grounds" for the truth values of all other sentences. But observe that we have not introduced any new metaphysical structure in doing so, beyond the metaphysical structure required to get the basic relations of "semantic guaranteeing" going in the first place. One upshot, then, is that Rosen's "clearest example" isn't really so clear, after all.

Finally, it is worth recalling that our discussion shows that one *really* subversive option is available, for the philosopher who doubts that "ground" picks out some metaphysically significant structure that subserves a distinctively metaphysical kind of explanation. After all, talk of what grounds what can just as easily be used to convey a position on what *good grounds* there are for drawing a distinction in a certain way—and those good grounds need involve no invocation of metaphysical structure.

Let's take stock. To me, perhaps the most important lesson to be extracted from the discussion so far concerns the extraordinary temptations of metaphysics. It can, I think, be very difficult to wean oneself from the habit of thinking in robustly realist terms. To be sure, sometimes we *should* be robust realists. But not always; nor should this be the default position to take once we have convinced ourselves that some concept X deserves to be part of our first-class philosophical vocabulary. And succumbing to this temptation can make it difficult to appreciate the resources of the respectful deflationist alternative. In the best of cases, that alternative can fashion a non-metaphysical vindication of the target concept X, by showing how the possession of a concept that functions in the way X does makes it possible for us to achieve some deep, pervasive, valuable goal of inquiry—where this goal in turn requires no special metaphysics for its characterization. You will misunderstand what such a respectful deflationist has achieved if you insist on warping her view into a thesis about what *metaphysically explains* X, when what she is in fact offering is an explanation of the *decisive reasons in favor of adopting* X, and using it in a certain way to structure inquiry. And you will likewise have been tempted into confusion if you think that she has anything to apologize for, should those decisive reasons advert to the distinctively and idiosyncratically *human* character of our goals.

References

Albert, D. (2012). Physics and chance. In Albert, D., Ben-Menahem, Y., & Hemmo, M., eds., *Probability in physics,* Springer, 17–40. Reprinted in Albert 2015; page references are to this latter printing.

Armstrong, D. M. (1983). *What is a law of nature?* Cambridge University Press.

Armstrong, D. M. (2001). Going through the open door again: Counterfactual versus singularist theories of causation. In Preyer, G., ed., *Reality and Humean supervenience: Essays on the philosophy of David Lewis.* Rowman & Littlefield, 163–76.

Beebee, H. (2000). The non-governing conception of laws of nature. *Philosophy and Phenomenological Research, 61*(3), 571–94.

Cohen, J. & Callender, C. (2009). A better best system account of lawhood. *Philosophical Studies, 145*(1), 1–34.

Fine, K. (2012). Guide to Ground. In Correia & Schnieder, eds., *Metaphysical grounding.* Cambridge University Press, 37–80.

Hall, N. (2007). Humean reductionism about laws of nature. Ms.

Hicks, M. (2018). Dynamic Humeanism. *The British Journal for the Philosophy of Science, 69*(4), 983–1007.

Hicks, M., & van Elswyk, P. (2015). Humean laws and circular explanation. *Philosophical Studies, 172*(2), 433–43.

Lange, M. (2013). Grounding, scientific explanation, and Humean laws. *Philosophical Studies, 164*(1), 255–61.

Lange, M. (2018). Transitivity, self-explanation, and the explanatory circularity argument against Humean accounts of natural law. *Synthese, 195*(3), 1337–53.

Lewis, D. (1983). New work for a theory of universals. *Australasian Journal of Philosophy, 61*(4), 343–77.

Lewis, D. (1994). Humean supervenience debugged. *Mind, 103*(412), 473–90.

Loewer, B. (2007). Laws and natural properties. *Philosophical Topics, 35*(1/2), 313–28.

Loewer, B. (2012). Two accounts of laws and time. *Philosophical Studies, 160*(1), 115–37.

Marshall, D. (2015). Humean laws and explanation. *Philosophical Studies, 172*(12), 3145–65.

Maudlin, T. (2007a). *The metaphysics within physics*. Oxford University Press.

Maudlin, T. (2007b). A modest proposal concerning laws, counterfactuals, and explanation. In Maudlin, *The metaphysics within physics*. Oxford University Press.

Miller, E. (2015). Humean scientific explanation. *Philosophical Studies, 172*(5), 1311–32.

Norton, J. (2003). Causation as folk science. *Philosophers' Imprint, 3*(4), 1–22.

Rosen, G. (2010). Metaphysical dependence: Grounding and reduction. In Hale & Hoffmann, eds., *Modality: Metaphysics, logic, and epistemology*. Oxford University Press, 109–36.

Russell, B. (1913). On the notion of cause. *Proceedings of the Aristotelian Society*, 13, 1–26.

Schaffer, J. (2009). On what grounds what. In Manley, Chalmers, & Wasserman, eds., *Metametaphysics: New Essays on the foundations of ontology*. Oxford University Press, 347–83.

Sider, T. (2011). *Writing the book of the world*. Oxford University Press.

van Fraassen, B. C. (1989). *Laws and symmetry*. Oxford University Press.

6
Pragmatism and the Challenge of Scientific (Dis)unification

David Danks

6.1 Pragmatist Philosophy of Science vs. Philosophy of Pragmatist Science

The term "pragmatism" has experienced a resurgence in the philosophy of science in recent years, providing a label for a wide variety of theories, methods, and ideas. Many claims are made about how pragmatist approaches lead to something different in the philosophy, methodology, or metaphysics of science, though that "something" varies quite substantially from author to author, or from scientific domain to domain, or from problem to problem. As the other chapters in this book show, there are certain similarities in much of what falls under the label of "pragmatist philosophy of science," but also significant diversity. I thus begin by thinking more generally about the intersection of pragmatism and philosophy of science, as it is sometimes unclear how we ought to evaluate pragmatist proposals in the philosophy of science. Following these initial remarks, I turn to the more specific focus of this chapter on the connections between pragmatist approaches and the unifiability or unification of scientific theories.

At a high level, I understand "pragmatism" to refer to a collection of methods and attitudes, rather than any particular content. More specifically, a pragmatist approach to, or theory of, *X* is one that focuses on the practical functions and impacts of *X*, as well as the functional needs of users of *X*. Pragmatist approaches find value in the utility of *X*, rather than solely in "truth," "verisimilitude," or some other, typically more metaphysical, notion. Of course, *X* might be useful partly because it is true, but on a pragmatist approach, that truth does not thereby provide intrinsic value for *X*, only indirect value by way of more accurate predictions, finer-grained control, or

David Danks, *Pragmatism and the Challenge of Scientific (Dis)unification* In: *The Pragmatist Challenge: Pragmatist Metaphysics for Philosophy of Science.* Edited by: H. K. Andersen and Sandra D. Mitchell, Oxford University Press.
DOI: 10.1093/oso/9780198805458.003.0006

other pragmatic virtues. I am deliberately quite vague in my description of pragmatism, precisely because there is arguably no single, clear formulation that can cover the disparate uses of the term. However, even these high-level comments indicate that pragmatism is more of a disposition, approach, or method, rather than any particular concrete content or commitments. Of course, those shared methods and questions will often lead to commonalities or similarities across various pragmatist accounts, but those are not thereby constitutive of "pragmatist theory of *X*."

Given these observations about the importance of practical impacts and needs, one can immediately see, as suggested by the title of this section, that there are two different places where the modifier "pragmatist" can be independently introduced into philosophy of science. That is, two different kinds of inquiry could arise at the intersection of pragmatism and philosophy of science. First, one could pursue a "philosophy of pragmatist science," in which one does relatively "realist" philosophy of science, but where "science" is characterized as a deeply pragmatist (for the scientists) endeavor. This approach is realist in the sense that the philosopher's efforts to describe, analyze, critique, and justify scientific theories and practices are judged by whether her account actually tracks the "facts on the ground"; does the philosopher accurately capture the way that the science *really* works or should work, or the *real* commitments of the scientists, or other aspects of what is *really* going on?

In this approach, the term "pragmatist" is applied to the science in the sense that the targets of inquiry—scientific theories and practices in some domain—are (the philosopher argues) themselves appropriately judged by pragmatist standards. For example, one might argue that scientific theories are accepted or disseminated based on their ability to provide experimental control, rather than truth. In this example, "pragmatist" applies to the scientific practices, but not to the philosophical claim about the scientific practice; the latter claim is intended in a realist way, as the philosopher thinks that she is capturing something about how science actually works. In practice, a large fraction, and perhaps even most, "pragmatist philosophy of science" is actually "(realist) philosophy of pragmatic science": the philosophical descriptions are argued to accurately describe scientific theories and practices, where the science is itself judged (internally) according to pragmatist standards of utility and benefit.

A second approach is to pursue "pragmatist philosophy of science," in the sense of judging the philosophy itself using pragmatist norms. In this tradition, one asks whether the philosophical approach or theory helps the scientists to achieve their goals, even if the philosophical account does not necessarily accurately describe the scientific world (in a realist sense). Importantly, the goals of the scientists need not be restricted to traditional scientific goals, but could include ambitions to serve the public good, or their own careers, or some other extra-scientific purposes (Kitcher 2001). Moreover, a scientist's goals need not always be the proximal psychological motivation or cause of her actions; one's goals may emerge or be clear only given a long(er) sequence of actions. The pragmatist utility of our philosophical theories and ideas for these scientific goals *might* depend on whether those ideas accurately capture the relevant scientific theories and practices. But critically, the truth or accuracy (or lack thereof) of the philosophical account is not directly relevant, but matters only to the extent that truth or accuracy is a precondition for helping the scientists to do their jobs more successfully.

Much of the philosophical work on scientific methodology arguably falls into this latter category, even though it is not often described as "pragmatist philosophy of science." Nonetheless, methodological analyses and recommendations are judged principally by the extent to which they help scientists achieve their goals. For instance, a methodological critique that provides no positive guidance or assistance is not pragmatically useful, and also often judged to be a relatively weak type of philosophy of science. Moreover, as the case of methodological analyses suggests, this kind of pragmatist philosophy often regards the scientists themselves as realists attempting to represent or mirror the world. Methodological guidance is frequently framed in terms of methods to better describe or understand the world, not "merely" to better control it or achieve other pragmatic goal.

More descriptive titles for these two projects—albeit, not the terminology that is commonly employed—would thus be "realist philosophy of pragmatist science" and "pragmatist philosophy of realist science," respectively. On this way of framing matters, "traditional" philosophy of science (inasmuch as such a thing exists) is usually "realist philosophy of realist science," trying to describe, analyze, and understand the actual content and practices of science, where those are understood as part of a truth-seeking enterprise (though one that might not actually be successful at finding truths). At the same time, we could have "pragmatist philosophy of pragmatist science," an endeavor that simply aims to help scientists succeed at their pragmatic goals; I return to this possibility in the final section, since there are arguably few (if any) examples of this type of work.

This initial discussion might seem to be unnecessary metaphilosophical commentary, but I suggest that it is important for understanding and evaluating efforts within "pragmatist philosophy of science." In particular, that label risks blending together two very different projects[1]—realist philosophy of pragmatist science (i.e., the subject of many chapters in this volume) and pragmatist philosophy of realist science (i.e., much methodological work)—that give quite different answers for the seemingly simple focal question of this chapter: what commitments or expectations, if any, should a philosopher of science who espouses "pragmatism" have about the likely unifiability of scientific theories? More colloquially: is pragmatism (in the philosophy of science) inconsistent with unified or unifiable science?

Although this question has not received extensive examination, the default answer for many people seems to be "yes, pragmatism and unification are problematically inconsistent." Section 6.2 explores the connections that seem to be widely perceived or assumed between pragmatism and disunification. Those arguments lead to Section 6.3, which provides a clearer understanding of "unification" or "unifiability" that is not grounded in the overly restrictive notion of "reduction." In particular, I argue that the unifiability of two scientific theories depends on the constraints that they imply for each other, where those constraints depend partly on a broad set of theoretical commitments. Given this understanding of "unification," I argue in Section 6.4 that the potential connections between (dis)unification and pragmatism actually depend on the nature of the pragmatism: in particular, we should expect disunification only if the pragmatism is accompanied by a drive for reification of scientific objects.

Realist philosophers of pragmatist science are unlikely to have a "reification drive" with respect to scientific objects, as they understand scientific theories to be evaluated by success criteria other than (or, not solely by) correspondence with reality. In contrast, pragmatist philosophers of realist science—scientific methodologists and others whose goal is to help truth-seeking scientists succeed—plausibly inherit such a "reification drive" from the scientists themselves. These researchers are likely to find they cannot construct coherent "pictures of the world" from diverse scientific theories. In other words, realist philosophers of pragmatist science are likely to have all the unification that they might want, while their counterparts will face a challenging type of disunification, precisely because they want and need

[1] Thanks to Clark Glymour for emphasizing (in multiple conversations) this problematic terminological ambiguity.

more from scientific theories than those can give. There are (at least) two different claimants to the term "pragmatist philosophy of science," but only one faces challenges from the potential (dis)unification of science.

6.2 Pragmatist Science and (Dis)unified Science

There is deep and widely held support for the existence of *some* connection between disunification and pragmatism about scientific theories and practice. Of course, such "support" is a bit difficult to measure, as there are as many varieties of pragmatism as there are pragmatists. Nonetheless, there seems to be relatively widespread agreement that pragmatist approaches—as with pluralist, antireductionist, and other "not garden variety realist" positions—inevitably result in disunified scientific theories, or at least, that disunified theories are a natural end-state for any pragmatist stance. Moreover, this agreement seems to be shared across both philosophers of science of all types (whether hardcore realist, deflationary realist, pragmatist, pluralist, perspectivist, or some other -ist); that is, all sides in the discussion seem to share the view that "other-than-realist" views[2] of scientific theories imply that scientific theories will, in practice, not be interestingly unifiable. Of course, the different sides in these debates draw different morals about the tenability of pragmatism from that particular conclusion. One argument of the present chapter is that both sides have accepted the putative conclusion far too quickly, but that argument must wait until Section 6.4. Here, we aim simply to understand the source and nature of the widespread belief that there is some tight connection between disunification and "other-than-realism."

At a (very) high level, the intuitive argument—again, seemingly accepted by all sides—starts with the observation that "other-than-realists" all agree that scientific claims, theories, and the like depend on, or are otherwise indexed by, features of the community, individual scientist, domain, measurement methods, and so forth. That is, all "other-than-realist" positions agree that some (perhaps all) scientific claims are shaped, influenced, or otherwise changed by factors that are not themselves modeled or

[2] I use this awkward term to encompass a wide swath of positions that all stand in opposition to the "traditional" realist, unificationist, reductionist picture of science. Of course, many of the proponents of those different philosophical positions would resist being grouped together given the significant differences in their positive claims.

represented within the science itself. As a result, diverse scientific domains and communities can produce significant diversity across these various types of scientific claims and theories, where that diversity is due to features that are not part of the particular scientific domains themselves.[3] Hence, unification will happen only by chance, if at all, as there is no theoretical reason to expect that the various other influences will cancel, balance, or otherwise not conflict in the ways that they change or shape the scientific claims. Moreover, we have descriptive, empirical reasons to expect disunification, as there are numerous case studies in which the science appears to be fragmented because of these factors (e.g., the many examples in Cartwright 1999; Danks 2015; Dupré 1993; Giere 1999; Kellert, Longino, and Waters 2006; Mitchell 2009; Mitchell and Gronenborn 2017; Wimsatt 2007; and so many more).

Standard realists and reductionists typically view this conclusion as a *reductio* of the "other-than-realist" views. If science (probably) cannot be unified, then it seems that science is incapable of providing a single, coherent view of the world. In particular, this conclusion suggests that many of our scientific theories are fundamentally inconsistent in some way. But since reality is presumably not inconsistent (whatever that might mean), "other-than-realism" seems to imply (from the realist's point of view) some fundamental failing in our scientific theories: science, as a collective activity, must be failing to track the truths about the world. At the same time, the realist can point to many explanatory, predictive, and experimental successes produced by efforts to unify scientific theories (Kitcher 1981; Bechtel and Richardson 2000; Sober 2003; Schouten and de Jong 2012), and thereby argue that we have abductive reasons to embrace the possibility of unification. The realist contends that such systematic advances are either accidental or miraculous for the "other-than-realist" positions. These other positions simply cannot (the realist argues) explain why a drive to unify is so frequently successful. As a result, for most realists, the implication of probably-necessary disunification is a deal-breaker: any philosophical account of science with that consequence is one that we should reject on both theoretical and pragmatic grounds.

In contrast, the "other-than-realists" seek to explain away or otherwise accommodate disunification, though the diversity of those -ists quickly

[3] Of course, some other science (e.g. sociology of science) might study those factors scientifically. The important point here is that, for example, biochemical theories might depend on factors that are not part of the domain of biochemistry.

becomes apparent in the diversity of their responses to this challenge.[4] One response is to argue that this disunification is methodological, not metaphysical (e.g., Mitchell 2009; Mitchell and Gronenborn 2017). This line focuses on the observation that different parts of the sciences use different methods and techniques, and so reveal different aspects of the world. That particular observation, however, does not imply metaphysical inconsistency, just as our use of different senses (which are sensitive to different aspects) to learn about an object in our environment does not lead to inconsistency or indeterminacy about that object. Certain types of limited or partial integrations are possible, despite the global disunification of science. A different response is to embrace metaphysical disunification (e.g., Dupré 1993), but to argue that this position is only problematic if we beg the question by presupposing a strongly essentialist picture of the world. More specifically, metaphysical disunification is problematic only if we try to have a global, integrated understanding of the world at a single moment in time, but we arguably never have such a goal (and might not even be capable of such a goal). Yet another response is to observe that science inevitably offers only partial models, and those abstractions or approximations are driven by factors outside of the particular scientific domain, such as our goals or instruments (e.g., Cartwright 1999; Danks 2015). The resulting disunification is principally epistemic or cognitive in nature, and so does not necessarily lead to any particularly troubling inconsistencies or failures.

My intention here is not to adjudicate whether disunification is a *reductio* of "other-than-realism," or simply an unproblematic side effect of it. Rather, I want to explore the largely unquestioned conclusion that they are intimately connected: is it actually the case that "other-than-realist" positions necessarily (or almost-surely) imply scientific disunification (of some sort)? Part of the challenge in answering this question is that, despite various suggestive connections between pragmatism and disunification, the relevant philosophical arguments are rarely clear about what it means for two different scientific theories to be unified, or to be unifiable. I thus turn now to the question of what it means for two scientific theories—more generally, two sets of scientific claims (that might not rise to the level of a full-blown theory)—to be unifiable or actually unified. Section 6.4 then argues that, on

[4] I paint with an admittedly broad brush here. All of the authors cited in this paragraph have significantly more complex and nuanced views than the ones I describe. My intention is only to point out different types of responses by the "other-than-realists," rather than to provide a careful exegesis of these sophisticated philosophical positions.

this understanding of unification, the almost-necessary connection is broken, though only for *some* philosophies of science. In particular, those who argue for pragmatist science can have all the unifiability and unification that they want, while those endorsing realist science face problematic disunification.

6.3 Unification, Constraints, and Commitments

I have so far used the terms "unification" and "disunification" as though they were clear and precise, but we must now explore them in more detail. One natural account of "unification," and the one that often seems to be implicitly presumed in these debates, is fundamentally reductionist about science. In particular, we might say that two scientific theories, T_1 and T_2, are unifiable if and only if there exists a scientific theory B (perhaps equal to T_1 or T_2) such that (i) T_1 and T_2 can be derived from B given auxiliary information Γ_1 and Γ_2, respectively (where each Γ can include background conditions, bridging principles, approximations, and so forth); and (ii) Γ_1 and Γ_2 are not inconsistent, whether because they are mutually compatible or restricted to distinct domains (or both). That is, the standard account presumes a kind of broad reductionism, and holds that two theories are unifiable just when they share a common reductionist base (regardless of whether we know that base). Given this starting point, we say that T_1 and T_2 are *actually* unified (not just unifiable) when we have identified appropriate B, Γ_1, Γ_2, as well as domains of reduction and the like. The difficulty of finding such a B will depend on details of T_1 and T_2; for example, unifying biochemistry and social psychology is probably much harder than finding a suitable B for theories of visual and auditory perception.

While this reductionist approach is intuitively appealing, it assumes that reductions to a common base are the only way to establish that two theories are unified or unifiable. However, reduction is not the only intertheoretic relation, and so this approach is overly restrictive. We can arguably demonstrate that two theories are compatible with one another (in the sense that they do not directly conflict) without necessarily identifying a shared reductive base B. More generally, reductions require that we establish a certain kind of equality or equivalence (whether syntactic, semantic, causal, or other) between B and T_1, and between B and T_2. While such equalities might be admirable goals in many contexts, they are also quite difficult—perhaps, unnecessarily difficult—to achieve. Instead, we can consider a broader view of unifiability and intertheoretic compatibility that allows for non-reductive connections.

Any particular set of theoretical commitments (perhaps a fully-fledged scientific theory, perhaps not) implies a set of constraints on other theoretical commitments, empirical implications, and so forth. As a trivial example, I might have the commitment that *X* and *Y* are correlated, perhaps because of empirical observations of such a correlation. In that case, my commitment implies the constraint that my other theoretical commitments about *X* and *Y* (whatever they might be) should predict such a correlation, or at least *not* predict *non*-correlation. This constraint is something that we ought not contradict or violate, at least to the best of our abilities and given our other constraints and commitments. More generally, we (*qua* scientists) have sets of theoretical commitments that might sometimes rise to the level of a scientific theory, but often will fall short. Nonetheless, these commitments imply sets of constraints that we ought (epistemically) endeavor not to violate. Moreover, these commitments do not map in a clean, one-to-one fashion onto claims in scientific papers. Most scientists have commitments that they do not articulate in writing, and many assertions in scientific papers are driven by convention (or responses to reviewers) rather than reflecting a deep theoretical commitment. In fact, the discovery and articulation of the *actual* theoretical commitments for a scientific community is frequently a substantive contribution of philosophy of science.

These commitments impose both soft and hard constraints. I have elsewhere defended a fuller theory of intertheoretic constraints (Danks 2014; see Thagard and Verbeurgt 1998 for a different account that is nonetheless consistent with the arguments here), but we do not need the full account here. Rather, we simply need the following three features: (a) constraints arise from commitments that might fall short of full theories; (b) constraints can be weaker and more indirect than what is required for a reduction; and (c) there can be different constraints (between the same sets of commitments) for different theoretical virtues. Point (c) is particularly relevant here, as it implies that, for example, a set of commitments might impose empirical or predictive constraints without thereby implying any significant ontological constraints. The "trivial example" from the previous paragraph provides such a case: the commitment to a significant *X*-*Y* correlation constrains the acceptable predictions of other theories, but does not constrain the objects that those other theories might posit, or the relationships that might obtain between those objects.

If we think about matters in terms of constraints, then we get an alternative account of unifiability. We start with a weaker notion: theoretical commitments C_1 and C_2 are *compatible* if and only if the intertheoretic constraints implied by C_1 are not inconsistent with the constraints implied by C_2. That is, two sets of theoretical commitments are compatible when they do not conflict, and so could potentially be simultaneously satisfied. This notion says only that C_1 and C_2 can be (rationally) believed at the same time. In contrast, the stronger notion of unifiability implies that C_1 and C_2 are, or could be, part of the same framework or system (in some sense). Thus, we can define: C_1 and C_2 are unifiable if and only if there exist (consistent) theoretical commitments U such that (a) the constraints implied by U are stronger (not necessarily strictly stronger) than the conjunction of constraints from C_1 and C_2; and (b) U is simpler (in some sense) than the conjunction of C_1 and C_2. Condition (b) is necessary for unifiability to be stronger than compatibility; without it, there is always the trivial "unification" $U = C_1 \cup C_2$ for compatible *C*s. Of course, that same condition implies that unifiability is no longer a binary predicate, but rather comes in degrees, corresponding to the relative simplicity of U. I contend that this is a feature, not a bug: two sets of theoretical commitments can be more-or-less unifiable depending on the extent to which there is a single, compact, shared basis for both.

This constraint-based account of unifiability provides a natural generalization of the reduction-based account with which we started this section. A reductive basis B (plus Γ_1 and Γ_2) provides an appropriate set of theoretical commitments for the ontological and predictive constraints of T_1 and T_2, and so those reduction-based unifications are a special case of the present framework. The constraint-based account has the added virtues of both not requiring theories but only theoretical commitments, and also employing a more nuanced view of intertheoretic relations. In many cases, we can find an appropriate U more easily than a reductive basis B, precisely because we do not necessarily need to establish an equivalence between, say, C_1 and U-plus-additional-claims. Continuing our trivial example, suppose that C_1 = "There is a significant X–Y correlation" and C_2 = "There is a significant Y–Z correlation." These are unifiable if we can find a set of commitments that imply the relevant constraints, which will almost surely be easier than finding a single, shared reductive basis that implies these claims. Moreover, this account fits with much of scientific practice: the drive for unification often takes the form of finding a "story" that accounts for everything we previously believed, even if that "story" does not provide strict reductions for the previous theoretical commitments.

6.4 Pragmatism, Reification, and Unifiability

We can now return to the overarching question of this chapter: if one is a pragmatist philosopher of science, should one expect that scientific claims or theories will be unifiable? The answer to this question can potentially both provide guidance about how to approach our philosophical analyses of particular scientific domains or problems, and also shape our broader understanding of scientific practices and communities. However, as argued in Section 6.1, this question is importantly ambiguous, as there are different ways to be a pragmatist in the philosophy of science. I contend that one's answer to the overarching question depends directly on one's resolution of this ambiguity: specifically, I argue in this section that philosophers of realist science should not expect unifiability while philosophers of pragmatist science should expect to have all of the unifiability that they want, and these both hold independently of whether one has a pragmatist or realist philosophical approach.

First, suppose that one is a philosopher of realist science, regardless of whether the philosophy is conducted in a realist manner (i.e., trying to understand what scientists are, or should be, actually doing) or in a pragmatist way (i.e., trying to help scientists be more successful in their truth-seeking endeavors, perhaps through methodological innovations). In these inquiries, the scientists are understood to be developing truth-conducive models of the world; scientific theories are, in some sense, supposed to be telling us how the world "really is." There are many different ways of filling out this core idea, but they all share the understanding that science is telling us about important features of the world. As a result, philosophy of realist science naturally leads to (attempted) reification of the objects postulated by the scientists themselves. Of course, this drive for reification can readily fail, since realist science need not succeed in the short run. Despite the scientists' best efforts, their theories and models of the world might fail to correctly identify and track the key (classes of) objects and properties. Nonetheless, the very idea of realist science is premised on a drive to identify and characterize the underlying nature of the world, and so philosophy of realist science naturally aims to understand our scientific theories as attempts to fulfill this drive, which requires that we (defeasibly) reify the contents of many scientific theories.

The problem is that we have good reason to doubt that reified scientific theories or models will actually be unifiable. The core challenge is the pragmatist's observation that there is not necessarily one single, privileged way to divide the world into objects and features. Rather, there are domains in which we seem to have multiple, incompatible sets of objects that are equally justifiable *by the realist scientist's own standards*. On the one hand, the realist scientist is committed to the "existence" of all these objects given the available-to-her standards. On the other hand, the incompatibility between the object-sets means that there can be no shared *U* that yields all of the reified objects.[5] Thus, the philosopher of realist science should not expect those sciences to be unified. Of course, the key to this conclusion is the existence of distinct sets of objects that are incompatible, yet equally justified by the realist scientist's own lights. I have elsewhere (Danks 2015) defended this claim in more detail; for space reasons, I provide only a high-level explanation here.

A (realist) scientific ontology would ideally be judged by its closeness-to-truth. The realist scientist aims to capture the way that the world "really is," and so a (proposed) ontology should be evaluated by how closely it corresponds to the underlying reality. However, we *qua* scientists do not possess an Archimedean point outside of the world from which to evaluate an ontology. Even if we developed a distance measure between a proposed ontology and the world, we have no independent access to one of the inputs to that measure (namely, how the world "really is"). Instead, scientific practice typically relies of converging indirect measures; there is no single "Discover the World" goal that can serve as the sole guide for scientific inquiry. We regard a proposed ontology as a good candidate for a realist science when it is appropriate or successful for diverse goals or challenges. For example, if a single ontology works well for prediction, experimental control, design of "soft" interventions or adjustments, and other tasks, then scientists typically conclude that they have (defeasible) reason to think that they are approaching "the" correct ontology.

This reasoning in favor of a single, reified, realist scientific ontology is essentially an instantiation of the intuitions behind the classic "no miracles"

[5] Unless she is prepared to adopt some kind of paraconsistent logic. Thanks to Liam Kofi Bright for pointing out this escape route for the realist.

argument. This argument strategy presupposes, however, that there is a single, unique ontology that will work best for all of the different goals and their accompanying measures, and we have both theoretical and empirical reasons to doubt this presupposition. For example, there are relatively ordinary systems for which we can prove that the "ideal ontology" (i.e., even in the limit of infinite data) for prediction given passive observations is incompatible with the ideal ontology for prediction given interventions or manipulations (see Danks 2015 for details about this and other cases). The realist scientist has to hope or assume that the "true" ontology is always ultimately the best on all relevant measures, but that hope is arguably mistaken.[6] And since we should expect incompatible ontologies depending on one's scientific goals, then we should expect disunification if we subsequently reify the relevant ontologies (as the philosopher of realist science is driven to do). More controversially, one might worry that the methodologist's pragmatic goal—helping the scientist to do her job better—is impossible if the scientist's job is understood in realist terms. One of the claimants to the label "pragmatist philosophy of science" arguably faces a fatal challenge from (dis)unification, though I consider a possible response in the next section.

One might instead be a philosopher of pragmatist science. In this case, I contend that disunification is not a problem. Pragmatist science is not, at its heart, concerned with understanding how the world "really is," but rather with developing systems—models, theories, ontologies, methods, and so forth—that satisfy various practical or pragmatic goals. Of course, those goals can be rather abstract (e.g., "prediction across diverse contexts"), but they do not include "Capture full truth" as a target (unless it happens to be a good way to achieve the other goals). As a result, there is little reason or motivation to engage in systematic reification of the objects postulated by pragmatist science. Although our scientific theories might appear to posit a rich, complex ontology, the pragmatist scientist is only committed to a type of functional or Ramsey-sentence-style characterization: there must be some reason for the success of the scientific theory and methods, but that reason need not be existence of the fully reified scientific ontology.

In the language of the previous section, the pragmatist scientist's theoretical commitments (that generate constraints) are only whatever is required

[6] At the same time, it is important not to overstate the conclusion of these arguments. In particular, they do not imply that "anything goes;" rather, there will typically be a close-to-unique ontology for each scientific goal/measure. The problem is just that those various ideal ontologies do not necessarily coincide.

for success on the pragmatic goals. Importantly, these commitments can extend significantly beyond "mere" empirical adequacy; I am certainly not suggesting that pragmatist science never (justifiably) goes beyond the phenomena. For example, successful prediction given interventions or manipulations implies that one's theory is tracking relevant parts of the underlying, unobserved causal structure. Alternately, the scientist might be committed to the scientific theory providing useful, informative explanations, which requires commitment to more than just the observable phenomena. At the same time, each of these commitments is tied to directly measurable goals, rather than the ill-defined (and unmeasurable) goal of "closeness to reality." As a result, the scientist's commitments are more limited, and are tied to the particular pragmatic goals and activities of the scientific community. Scientific objects, and the scientist's theoretic commitments about them, are no more, and also no less, than what is needed in order to secure the pragmatic success of the scientific theories and models.

Since our theoretic commitments are limited to only what is functionally required for the pragmatic goals, the core question is whether the constraints provided by those diverse pragmatic demands are compatible with one another.[7] Thus, I contend that we should expect that disparate scientific theories will be unifiable to the extent that the world is consistent across scientific investigations. As a simple example, notice that "simple" prediction is usually one of our pragmatic goals, and so passive observations frequently provide pragmatic constraints. Unifiability of theories derived from those observations will thereby depend partly on those observations not being contradictory. For instance, if our observations are directly contradictory (e.g., we find a positive *X*-*Y* correlation in one experiment and zero *X*-*Y* correlation in another), then we should not expect unifiability of theories that are based on those two different sets of observations. Of course, we presumably would never have expected unifiability in a case such as this one, since we have directly contradictory data. These are exactly the

[7] There are thus close similarities between my arguments and Sandra Mitchell's *integrative pluralism* (Mitchell 2009; Mitchell and Gronenborn 2017; and other writings), as we both emphasize the possibility of integration or unification of theories when we restrict matters to particular theoretical virtues, or particular purposes. For example, Mitchell and Gronenborn (2017) show how multiple models can be integrated to yield "predictive hypotheses, and...empirical models" (p. 705), both of which are tied to specific goals or virtues, rather than attempting to provide global, universal models. The views in this chapter (and related ones that I have articulated elsewhere, as in Danks 2015) do differ from integrative pluralism in various ways, which is why I have not framed my arguments explicitly in those terms. This chapter can, however, be read partly as trying to provide a more precise explication of some of the types of integration discussed by integrative pluralists.

situations in which we *qua* scientists try to either identify a relevant contextual difference (thereby eliminating the contradiction) or explain away one of the sets of observations as anomalous in some way. Nonetheless, it is helpful to see a simple instance of how the world might fail to be consistent across scientific investigations.

More generally, though, I contend that we should typically expect appropriate kinds of consistency, precisely because of the more limited nature of those pragmatic constraints. The pragmatist scientist can be understood as a certain kind of minimalist, as she does not commit to more (from her theories) than is required to account for her pragmatic successes. Again, I want to emphasize that I am not conflating pragmatist science with some type of instrumentalism or behaviorism; scientific successes can require commitments that go far beyond the observational and experimental data. Nonetheless, the pragmatist scientist is not committed to anything beyond "whatever is necessary for the practical successes," and so non-unifiability would require that there be types of practical successes that cannot be mutually achieved using the theories.[8] And I contend that different types of pragmatic goals should not be incompatible unless the world itself is incompatible between contexts, precisely because pragmatic goals are ultimately grounded in (though not constituted by) measurements and other "direct" engagements with the world.[9]

In particular, since the pragmatist scientist does not automatically reify the objects of her theories, we cannot infer non-unifiability solely from incompatible objects (in the different theories). For example, I earlier argued that different ontologies could be appropriate for the goals of prediction given passive observation versus prediction given experimental manipulations. This observation is unproblematic for the pragmatist scientist, as these "ontologies" are really just different ways of regarding the same system to achieve different practical successes. Pragmatism (and other other-than-realisms) need not lead to a problematic pluralism. Nothing about those

[8] Of course, there is a trivial sense in which particular practical successes might be incompatible: I cannot, for example, simultaneously satisfy the goals of "paint my house entirely red" and "paint my house entirely blue." However, these two goals are clearly of the same type, and so we should not expect simultaneous satisfaction. This observation does, however, raise interesting questions about how to individuate different pragmatic goals. In Danks (2015), I suggested that the notion of a loss function (as used in optimization theory and machine learning) might provide some guidance, though that clearly is not the whole story (see also Wellen and Danks 2016). Unfortunately, space constraints preclude a detailed examination of types of pragmatic (scientific) goals.

[9] I am obviously assuming here that we are concerned only with unifiability of "reasonably successful" theories (where that modifier can be left appropriately vague).

particular pragmatic goals requires the scientist to reify the objects in either theory. Other pragmatic goals might require such reification, but presumably only given stronger, more wide-ranging commitments and measurements.[10] Again, the key is to ask "what are the scientist's commitments for each theory?" For the pragmatist scientist, those will typically be significantly more minimal and better grounded in the world (compared to a corresponding realist scientist), and so they can all be simultaneously satisfied. That is, her scientific theories provide complex "inference tickets," not pictures or mirrors of the world, and so there is no particular incompatibility or non-unifiability. Of course, this pragmatist scientist also draws much more minimal conclusions—she does not "know" as much about the world—but I suggest that is entirely appropriate: she asserts only what is required by her various theoretic and pragmatic commitments and constraints, rather than (over)extending to a highly reified and realist picture.

If we now return to the claims and arguments of Section 6.2, we can see that the supposedly tight connection between "other-than-realism" and (problematic) disunification holds only for the philosopher of realist science, and not the philosopher of pragmatist science. That is, the debates involved a significant amount of people talking past one another due to different fundamental presuppositions about the nature and goals of science. Suppose one is a philosopher of realist science. In that case, the arguments offered in favor of "pragmatist science" will naturally seem to imply non-unifiability, but only because one is independently committed to science being a realist enterprise, including the (hopeful) reification of objects in scientific theories. Of course, the philosopher of pragmatist science rejects exactly this independent commitment: she holds (or *should* hold) that scientists are committed only to whatever is necessary to account for their pragmatic successes. And those more minimal theoretic commitments (for multiple scientific theories) will typically be unifiable, precisely because the commitments are weaker than the philosopher of realist science desires. The philosopher of pragmatist science has all of the unifiability that she could want, but that is not enough for the philosopher of realist science. The "debate" is not really about the implications of disunification after all, but

[10] As an aside, this approach suggests a different way of thinking about the classic No Miracles Arguments (NMA), as an NMA *also* requires demonstration that the collection of practical successes (that form the initial premises of that NMA) could only be explained through a reification commitment toward the relevant theory or theories. "Mere" successes are insufficient to ground an NMA on this view; rather, we have to do additional argumentative work that engages with the particular pragmatic goals and contexts.

rather about different conceptions or expectations of the unification that scientists need.

6.5 Toward a Pragmatist Philosophy of Pragmatist Science?

This chapter has not directly argued that we should be philosophers of pragmatist science, but I do believe that the arguments here lend indirect support for that stance. Epistemic humility suggests that we ought not commit ourselves to more than is required to explain our successes in reaching our goals (including prediction, explanation, control, and so forth). Philosophy of pragmatist science has exactly this kind of minimal, humble stance: theoretic commitments do not automatically extend to reification of the objects. Moreover, philosophy of pragmatist science can help to provide us with a "middle ground" (between the extremes of realist science and behaviorist science) that can handle or explain away many concerns about one or the other extreme (see also many of the chapters in this volume). Although my arguments about unifiability provide only indirect support for philosophy of pragmatist science, I nonetheless want to close by taking it as a starting point, and asking speculatively about whether we should be realist or pragmatist philosophers of pragmatist science. That is, rather than asking about the nature of science, I finish by asking about the possible natures of our philosophy. In particular, if my earlier arguments are correct that much of standard scientific methodology (i.e., pragmatist philosophy of realist science) faces a significant challenge from (dis)unification, then could we instead apply its pragmatist approach to pragmatist science?

Realist philosophy of pragmatist science is relatively straightforward to understand, as it involves "ordinary" descriptive and normative analyses of science, understood in terms of its pragmatic (not realist) goals. As I argued earlier, much of the work labeled "pragmatist philosophy of science" is best-understood as realist philosophy of pragmatist science, as the authors aim to produce accurate (i.e., "truthful") characterizations of the way that (pragmatist) science does or should proceed. Interestingly, much of the work in this vein has not turned its arguments for pragmatism on itself: why, for example, should we think that philosophical methods are more truth- or accuracy-conducive than scientific methods (which these authors often argue are focused on other goals or purposes)? Regardless, the general project and evaluation standards of realist philosophy of pragmatist science are relatively clear, and it has undoubtedly led to a more sophisticated and nuanced understanding of the nature of much of scientific practice.

Pragmatist philosophy of pragmatist science is a stranger beast, as the evaluation standards are quite different. I earlier argued that pragmatist philosophy of *realist* science is currently largely composed of methodological work, developing novel methods—experimental, observational, analytic, and so forth—and then testing them to determine their reliability or accuracy in relevant conditions. For example, much of the machine learning research on causal discovery methods (e.g., Pearl 2000; Spirtes, Glymour, and Scheines 2000; Spirtes 2010; Malinsky and Danks 2018) is, I suggest, naturally understood as pragmatist philosophy of realist science. That research area involves analyses of science as a truth-seeking enterprise (so is "philosophy of realist science"), but where those analyses are evaluated based on whether they contribute directly to the goals of science (hence "pragmatist philosophy"). Importantly, the evaluation in terms of "contribution" does *not* actually require that the methods be reliable; we only need to ask whether positing the method improves the scientists' abilities to reach their (truth-seeking) goals. Of course, one natural way to help the scientists is to provide methods that actually are reliable, but accuracy or reliability is not a necessary condition for helping (truth-seeking) scientists. The development of a new method that is faster-but-less-accurate might, for example, improve scientists' ability to discover the truth, depending on other constraints that they face. More generally, the important point is that pragmatist philosophy of science should ultimately be judged by how it improves or advances the science, rather than whether the philosophical analyses themselves conform to a more traditional (descriptive or normative) truth-seeking standard.

We can thus see that pragmatist philosophy of pragmatist science would involve, in part, attempts to provide methods, techniques, and analyses that advance science as a pragmatist, not-necessarily-truth-seeking enterprise, and where those proposals would be judged purely by whether they help the scientists to achieve their goals. When framed in this way, it is unclear whether there is much, if any, pragmatist philosophy of pragmatist science at the current time (though Bright, Malinsky, and Thompson 2016 suggest such a project in several places). An example of this type of work would be a novel methodological proposal that is evaluated based on whether it helps scientists achieve their (non-truth-seeking) pragmatist goals such as influencing policy to advance human flourishing. As noted earlier, though, new methodologies are almost always evaluated based on their reliability or truth-seeking capabilities (i.e., using standards of realist science). Part of the challenge here is that we do not necessarily have a clear catalog or taxonomy of scientists' pragmatic goals, and so it is not quite clear what needs to be "advanced" by such suggestions. Moreover, much of the rhetoric around science presents it as truth-seeking, and so there would be natural challenges to proposing a new method that does not purport to advance that putative goal. Nonetheless, even this simple hypothetical points toward possible analyses that broaden the scope of "philosophy of science" to include pragmatism throughout all of our efforts.

Acknowledgments

These ideas were originally presented at the "Philosophy of Science: The Pragmatic Alternative" conference, held in 2017 at the University of Pittsburgh Center for Philosophy of Science. Thanks to audience members, particularly Holly Andersen and Sandy Mitchell, for their thoughtful questions. Thanks to Holly Andersen, Liam Kofi Bright, and Sandy Mitchell for comments on an earlier written version of this chapter. Most of this chapter was written while the author was on the faculty at Carnegie Mellon University.

References

Bechtel, W., & Richardson, R. C. (2000). *Discovering complexity: Decomposition and localization as strategies in scientific research*. Princeton University Press.

Bright, L. K., Malinsky, D., & Thompson, M. (2016). Causally interpreting intersectionality theory. *Philosophy of Science, 83*(1), 60–81.

Cartwright, N. (1999). *The dappled world: A study of the boundaries of science*. Cambridge University Press.

Danks, D. (2014). *Unifying the mind: Cognitive representations as graphical models*. The MIT Press.

Danks, D. (2015). Goal-dependence in (scientific) ontology. *Synthese, 192*(11), 3601–16.

Dupré, J. (1993). *The disorder of things: Metaphysical foundations of the disunity of science*. Harvard University Press.

Giere, R. N. (1999). *Science without laws*. University of Chicago Press.

Kellert, S. H., Longino, H. E., & Waters, C. K, eds. (2006). *Scientific pluralism*. Minneapolis: University of Minnesota Press.

Kitcher, P. (1981). Explanatory unification. *Philosophy of Science, 48*(4), 507–31.

Kitcher, P. (2001). *Science, truth, and democracy*. Oxford University Press.

Malinsky, D., & Danks, D. (2018). Causal discovery algorithms: A practical guide. *Philosophy Compass, 13*(1), e12470. doi:10.1111/phc3.12470.

Mitchell, S. D. (2009). *Unsimple truths: Science, complexity, and policy*. University of Chicago Press.

Mitchell, S. D., & Gronenborn, A. M. (2017). After fifty years, why are protein X-ray crystallographers still in business? *The British Journal for the Philosophy of Science*, 68(3), 703–23.

Pearl, J. (2000). *Causality: Models, reasoning, and inference*. Cambridge University Press.

Schouten, M. K. D., & de Jong, H. L., eds. (2012). *The matter of the mind: Philosophical essays on psychology, neuroscience and reduction*. Wiley-Blackwell.

Sober, E. (2003). Two uses of unification. In F. Stadler, ed., *The Vienna Circle and logical empiricism: Re-evaluation and future perspectives*. Kluwer, 205–16.

Spirtes, P. (2010). Introduction to causal inference. *Journal of Machine Learning Research, 11*(5), 1643–62.

Spirtes, P., Glymour, C., & Scheines, R. (2000). *Causation, prediction, and search* (2nd ed.). The MIT Press.

Thagard, P., & Verbeurgt, K. (1998). Coherence as constraint satisfaction. *Cognitive Science, 22*(1), 1–24.

Wellen, S., & Danks, D. (2016). Adaptively rational learning. *Minds & Machines, 26*(1), 87–102.

Wimsatt, W. C. (2007). *Re-engineering philosophy for limited beings: Piecewise approximations to reality*. Harvard University Press.

7
Pragmatism, Perennialism, and the Physics of Ignorance

Laura Ruetsche

Investigations of the foundations of quantum field theories have suggested (at least to me) the thesis that theory specification has a pragmatist dimension: strategies for equipping physical theories with content, if sensibly pursued, eventuate in contents indexed not only (or not just) to the way the world is, but also to our aims in using our theories and the circumstances we use them in. This chapter assesses a "fundamentalist" response to the move to pragmatize theoretical content. The apparent appeal of the move, the fundamentalist response claims, derives from transient artifacts of the present incomplete state of physics. Fundamental physics, the response continues, can be properly understood if only it's understood as representing the way the world is. Anyone who thinks otherwise has paid too much attention to incomplete and unfundamental sciences! Of course, none of the physics we have right now is genuinely fundamental. To directly assess the fundamentalist response, we need to know things of which we are ignorant—the future of science, the nature of fundamental physics. Here, I undertake an indirect assessment, by developing two reasons to predict that future scientific theories, including theories of "fundamental physics," will continue to be best understood as possessing pragmatized content.[1]

[1] For desperately needed feedback on early drafts, I am obliged to Holly Andersen, David Freeborn, Christopher Mitsch, Paul Teller, Jingyi Wu, and participants in the 2017 PragMaps workshop at the University of Pittsburgh.

Laura Ruetsche, *Pragmatism, Perennialism, and the Physics of Ignorance* In: *The Pragmatist Challenge: Pragmatist Metaphysics for Philosophy of Science*. Edited by: H. K. Andersen and Sandra D. Mitchell, Oxford University Press.
 DOI: 10.1093/oso/9780198805458.003.0007

7.1 Introduction

Working on the foundations of quantum field theory (QFT) has convinced me that the best way to make sense of a successful physical theory is *not* to afflict upon it a single fixed interpretation that equips it once-and-for-all with an univocal content presenting the theory's picture of *the* way the world is. The best way to make sense of a physical theory is rather to associate with that theory a galaxy of interpretations, with different interpretations supporting its application in different circumstances. Thus I support a pragmatist thesis about the interpretation of physical theories. That thesis is: strategies for equipping physical theories with content, if sensibly pursued, eventuate in contents indexed not only (or not just) to the way the world is, but also to our aims in using our theories and the circumstances we use them in (see Section 7.4 for more and Ruetsche 2011 for much more). Here I try to defend the thesis against the criticism that its appeal, such as it is, derives from resting too much weight on the transient frailties of developing science. Really fundamental physics, the criticism goes, will admit a single, satisfying, and univocal interpretation that makes sense of all its applications. The content of really fundamental physics, the criticism concludes, lacks a pragmatic dimension.

The chapter proceeds as follows. Section 7.2 offers an account of what pragmatist approaches to philosophy of science *are*, that my view about the interpretation of physical theories counts as an instance of them. The account contrasts pragmatist approaches with what I call *perennialist* approaches, and along two axes. One axis concerns the expected shape of an adequate philosophical analysis; the other axis concerns methods appropriate to articulating an analysis of that shape. Section 7.3 argues that pragmatist *methods* honor an important methodological norm that perennialist methods flout. Pragmatist methods enjoin close attention to concrete details and specific applications. Paying such attention to members of a family of theories including QFT, Section 7.4 motivates the pragmatist thesis that the best way to understand those theories indexes their contents to contingent circumstances. Section 7.5 addresses the "Perennialist's Apology," which is the criticism that this and similar pragmatist theses are artifacts of the incomplete and imperfect condition of the bits of science pragmatism engages. The Perennialist's Apology rests on a prediction: the future of science will deliver fundamental theories unsullied by pragmatized content. Section 7.5 offers two reasons to bet against this prediction. Section 7.6 distills morals for perennialism, pragmatism, representation, and realism.

7.2 Pragmatism and Perennialism

One way into our topic, "pragmatic approaches to philosophy of science," is to ask: what other approaches are there?—and, how are pragmatic approaches different from these? (A theme to be developed presently is the methodological value of having alternatives in view.) A bit of mid-twentieth-century philosophical currency—"perennial philosophy"—affords a useful counterpoint:

> Of the philosophical phrases which have come into popular use during the XXth century, perhaps none is more curious than "perennial philosophy" or, in its more common Latin form *philosophy perennis*... *Philosophia perennis* is a philosophy which endures; its truth is considered to persist from generation to generation, long after ephemeral philosophical fads and fashions come and go. (Schmitt 1966, 505)

From the vantage of perennial philosophy, we're asking same questions Socrates did or a sentient alien might; these questions have (up to incidental local variations) the same answers; the answers reflect the way the world is. We may not yet have identified those answers, but they're what philosophy should be aiming for.

The perennialist's understanding of the *aim* of philosophy has implications for its *method.* The very possibility of perennial philosophy rests on the hope that the resources invoked in constructing answers to philosophical questions are available to all comers, regardless of their socio-historical location or peculiarities (e.g., "fads") pertaining thereto. The hope has a comforting consequence. If local contingencies don't matter to how a philosophical question is answered, philosophy is a project to which contingent ignorance is no impediment. What you don't know about how the world *actually* is can't hurt you, qua perennial philosopher.

What can help you? Resources available to all who inquire are resources appropriate for perennial philosophy. They might include (1) rationality (narrowly understood as an automatic endowment of inquiring creatures, not as something you grow into, given enough work undertaken in the right circumstances); (2) logic and mathematics; (3) metaphysical considerations (where these are fixed and universal). For investigations, like the present one, into science, another appropriate resource for the perennialist might be (4) methodological considerations, where these are anchored in a fixed and immutable Nature of Science.

It's noteworthy that (1) expresses a substantive view about the nature of rationality, and casting (2)–(4) as perennial resources presupposes substantive views about how rationality so construed functions in the fields of mathematics, metaphysics, logic, and science. For instance, metaphysical considerations are suitably general perennial resources if metaphysics consists in synthetic a priori statements which can be discerned by exercises of pure reason—but not if "metaphysics" is shorthand for a set of answers to Carnapian external questions. Put another way, (2)–(4) are expressions of the view that logic, mathematics, metaphysics, and philosophy of science are examples of perennial philosophy. These inquiries *can* be adeptly pursued without appeal to "adulterating" considerations that are artifacts of contexts of inquiry, and *if* successfully pursued will issue answers to philosophical questions that abstract away from such concreta. Although nothing's stopping the perennial philosopher from lavishing attention on messy contingencies, it follows from how she understands her endeavor that lavishing attention on messy contingencies will do nothing to promote it.

One way to hear "pragmatist" is as an alternative to "perennial." Even if the pragmatist philosopher poses questions similar to those posed by the perennial philosopher (and this remains to be seen), she countenances answers of a different shape, requiring for their construction resources unnecessary for perennial philosophy. The pragmatist entertains the possibility that answers to apparently perennial questions *could be* indexed to contingent circumstances contributing to the local conditions in which those questions are raised. And she has a correlate interest—one arguably motivating a variety of philosophical question absent from the perennialist's radar screen—in understanding how and why the indexing of answers to local conditions occurs. Thus the pragmatist philosopher of science puts features *not* common to all who inquire to philosophical use. Examples of these features include our aims and desires, the current state of technological and theoretical resources, what Longino calls contextual values—and all the ways considerations like the foregoing influence how we use theories.

Here are two examples contrasting perennial and pragmatist approaches to the philosophy of science:

- "Naturalistic" metaphysicians and Carnap on ontology. For Tim Maudlin,

  "Metaphysics is ontology. Ontology is the most generic study of what exists. Evidence for what exists, at least in the physical world, is provided solely by empirical research. Hence the proper object of most metaphysics is the careful analysis of *our best scientific* theories (and especially of fundamental physical theories) with the goal of determining what they imply about the constitution of the physical world." (Maudlin 2007, 104)

Maudlin casts "what exists?" as a perennial question, and urges us to obtain an enduring answer by means of "careful analysis" of fundamental physics. By contrast, Carnap insists that, in the happy circumstance that we encounter a successful scientific framework, "we have to face…an important question; but it is a practical, not a theoretical question; it is the question of whether or not to accept the new linguistic forms. The acceptance cannot be judged as being either true or false because it is not an assertion. It can only be judged as being more or less expedient, fruitful, conducive to the aim for which the language is intended. Judgments of this kind supply the motivation for the decision of accepting or rejecting the kind of entities." (Carnap 1950, 31). For Carnap, "what exists?" has different answers in different circumstances, answers which aren't aptly judged to be right or wrong, only more or less appropriate. Both the articulation and the assessment of answers requires attention to local contingencies.

- Mackie and Woodward on causation. On the perennial hand, we have Mackie's once-and-for-all analysis of an event's cause as an insufficient but non-redundant part of an unnecessary but sufficient condition for that event. On the pragmatist hand, we have Woodward's descriptive cum evaluative investigation of what sorts of causal reasoning we engage in, with what aims, and also (given the foregoing) what methodologies are appropriate ones for causal reasoning (Woodward 2023). (Cause is a concept running through a number of sciences. But similar contrasts can be drawn with respect to concepts central to specific sciences—for instance, gene or species or mass or pathology.)

These perennialist/pragmatist pairs dramatize a contrast between *what theories represent* and *how we use theories*, a contrast central to how many pragmatists understand their projects.[2] Huw Price distinguishes between what we might call a metaphysical stance with respect to a vocabulary—a stance which takes the primary question to be whether the claims distinctive of the vocabulary are *true*—and a genealogical or anthropological stance, which is interested in why creatures like us come to employ the

[2] E.g. Teller (2012) issues a plea, inspired by the consideration of scientific practice, to move beyond an old-fashioned notion of truth conceived in terms of literal, precise, and accurate representation.

vocabulary to begin with (2009, 25)—and chronicles the perils of adopting the former. Woodward also sees an alliance between pragmatist approaches and skepticism about representation:

> Historically, many pragmatists have been skeptical about (or hostile to) 'representation' as a useful concept for understanding how language and other sorts of activities such as scientific theorizing works, at least when representation is understood, as it often has been, in terms of notions like mirroring, picturing, correspondence and the like. (2023, 9)

I however will try to resist skepticism about representation, at least in its starkest form. I don't take "what theories represent" and "what theories are for" to lie on opposite sides of a methodological fault line. Focus on one aspect needn't preclude focus on the other. After all, a theory's representational capacities could have *something* to do with its patterns of and aptness for use, and our aims in applying it could shape its representational capacities! Pragmatists can care about representations, too, provided they express that care by attention to the uses to which representations are put. What is more, paying attention to representation may give pragmatists and perennialists (avowedly concerned with representation) common points of reference. And this could help us all navigate the tangle of considerations for and against pragmatist and perennial approaches to science.

It's important to distinguish *pragmatist methods* from *pragmatist theses* about the upshots of those methods. Pragmatist methods pay philosophically informed attention to local circumstances and concrete examples. Pragmatist theses concern the results of such investigation and contend that local contingencies matter substantially to philosophical analyses. Consider a putatively perennial analysandum *X*. A pragmatist thesis concerning *X* is that analyses of *X* keyed to the various habitats in which *X* functions defy a common perennial analysis. By contrast, the perennial philosopher expects a univocal analysis untethered to local accidents, and accordingly sees little cost in pitching her investigations at Theory *T*, or laws of nature as such, or science conceived in full abstraction. For she regards variations in concrete instantiations of each of these abstractions to be devoid of philosophical interest. A metaphor capturing this methodological difference depicts the perennial philosopher surveying an expanse of garden plots from the lofty heights of a hot-air balloon, a perspective which can obscure ground-level differences between those plots. The pragmatist, by contrast, is covered with dirt and down in the weeds.[3]

If pragmatist theses are true, *pluralism* is an anticipated consequence of adhering to pragmatist methods. Against the backdrop of perennial expectations that concept *X* is unitary, if accounts of *X* indexed to different contexts assume forms that defy a common analysis, pluralism about *X* results.[4] However, and this will be important soon, pluralism is not a *necessary* consequence of applying methods of inquiry appropriate to pragmatism. For it's possible that for some values of *X*, the many analyses crafted of *X* in its many natural habitats share a core structure that passes muster as a perennial analysis of *X*, that is, an account unadulterated by contextual considerations, of what *X* is, an account of enduring validity. In this case, following the pragmatist methods of examining concrete cases results in a unitary analysis congenial to the perennialist.

Obviously it would by my lights be ironic if the analysandum "pragmatism" admitted a unitary analysis. I don't mean the foregoing as a definitive account of conditions necessary and sufficient for philosophical approaches to count as instances of pragmatism. I offer it because it both suits and explicates my present purposes. One of these, prosecuted in the next section, is to argue that pragmatist methods adhere to a norm that perennialist methods flout.

7.3 In Praise of Pragmatist Method

Here's a suggestion, foregrounded in some of the feminist science literature but broadly relevant to the epistemology of science: in science, there's an epistemic value that operates at the level of methodology. In a slogan, the value is: don't silence evidence. In more detail, the value exhorts the scientist investigating a space of hypotheses to pursue protocols that enable the emergence of a wide range of evidence relevant to distinguishing between those hypotheses. An example of an experimental protocol defying this

[3] In a friendly dig at Arthur Fine, Ruetsche (2015) calls these grubby engagements with ground-level plots "locavore" methodologies.

[4] Some examples: for pluralism about genes, see Waters (2006); for pluralism about the ontology of biology, see Mitchell (2003). It's a bookkeeping matter whether to count this sort of fragmentation as a single concept admitting multiple analyses or as the discovery that what we regarded to be a single concept turned out instead to be many concepts. However, the books are kept, the philosophical surprise for the perennialist remains.

value might be an investigation of gender experiences that uses a survey instrument requiring respondents to identify themselves as either men or women (with no other option). Such a protocol makes it difficult for evidence of human experiences that don't fit neatly into that dichotomy to register. But those experiences are evidence relevant to hypotheses about gender experience.

Anderson (2004) is a sustained discussion of another example: research addressing whether divorce is damaging for women. Researchers adhering to an experimental protocol employing "objective" gauges of well-being, such as household income, concluded that divorce is damaging, citing such evidence as the lower household income divorced women typically have compared to the household incomes they had when they were married. A salient consideration obscured by the researchers' protocol is the possibility that less money could conduce to greater well-being if after divorce women have *more control* over their household income. Although this possibility is instantiated by many women after divorce, it emerges only through research protocols (like those followed by a team of investigators led by the psychologist Abby Stewart) that broaden the conception of well-being in way that allows research subjects to contribute testimony about the impact on that commodity of "objective" measures. Dubbing the epistemic virtue upheld by Stewart's team "fruitfulness," Anderson explains why it matters:

> One research design is more fruitful than another, with respect to a controversy, if it is more likely to uncover evidence supporting (or undermining) all, or a wider range of sides of the controversy.... For example, the conception of divorce as loss, presupposing a negative evaluation of divorce, will be able to guide research toward discovering the negative but not the positive features of divorce. By comparison, the Stewart team's value-laden conception of divorce as involving both loss and opportunities for growth is more epistemically fruitful, relative to controversies about the overall value of divorce, in that it allows us to uncover evidence bearing on both the pros and the cons of divorce. (2004, 20)

One might worry that the fruitfulness norm has limited bite insofar as "relevance" is keyed to hypotheses under investigation. Often, failures to meet a fruitfulness-like methodological norm—failures to register salient evidence—are only visible as such in retrospect, once hypotheses emerge which might have been supported by evidence suppressed by the protocols failing to meet the norm (see e.g., Hrdy 1986). That doesn't make the norm totally toothless in practice, though—there's no practical impediment to using the norm to fault protocols suppressing evidence for *hypotheses already on the table.*

Pragmatist and perennialist approaches to the philosophy of science are already on the table, as are their allied research designs of (respectively) grubbing in the dirt and lofty aerial survey. The fruitfulness norm has traction. And it favors the pragmatist research design, which enjoins close attention to ground-level practices. It is only through such attention that philosophically salient differences between those practices—evidence supporting pragmatist theses—might come into focus. But so might emerge evidence supporting the sorts of univocal analyses anticipated by the perennialist. Perfectly general accounts should, after all, apply to all instances. And so ground-level investigation of specific theories stands to confirm perennial accounts. Pragmatist methods are fruitful. Perennialist methods, however, are not: if there are insights to be had at the ground level which are not accessible from the hot air balloon, engaging science only from the perspective of the hot air balloon blocks the entry of those insights into evidence. Lofty disregard for specifics is a research design that violates the fruitfulness norm. Fortunately, for the perennialist, it's also a research design that's optional. Given her aims, she has no reason to get down in the weeds. But it's not incoherent for her to do so.

7.4 A Case for Content Pragmatized

7.4.1 The Perennialist's Apology

Adhering to fruitful pragmatist research designs, I'm going to examine some concrete bits of science afforded by linear and interacting QFTs, hailed by many as "fundamental" physics. Fundamental physics merits examination in this context due to a special role it often plays in debates over pragmatist and perennial approaches. That role is to underwrite a dialectical maneuver I'll call "the Perennialist's Apology." I'll explicate the Apology before turning to the concrete examination of QFTs.

"Fundamental" physics is so-called because it is imagined to be a complete and autonomous guide to what the physical world is at base like (cf. the quotation from Maudlin on p. 183). No legitimate question of physics is left unanswered by fundamental physics, and there are no exogenous empirical facts in virtue of which fundamental physics governs as it governs. Properly understood, fundamental physics discloses the whole truth about the physical world. Fundamental physics may be the last bastion from which perennial philosophers might defend their aim at univocal and abiding philosophical analyses from pragmatist affronts. Pragmatists confront the perennialist with case studies supporting pragmatist theses to the effect that analysands (gene, or aggression, or cause) should be understood in different ways in different contexts. In response, perennialists might issue

The Perennialist's Apology. *Of course*, you're going to encounter suggestive messiness if you fixate on evolutionary biology or causal reasoning in the social sciences. Those disciplines are not only superficial hodgepodges but also superficial hodgepodges rife with direct or analogical significance for the human condition—so hodge-podges of exactly the sort upon which one would expect "pragmatist" considerations to intrude. In those disciplines there is both opportunity and motive for such intrusion. There is opportunity because there is no fundamental way the biological or social world is, leaving room for interests, goals, and so on, to shape the way we take those worlds to be. There's motive because those are *our* interests, goals, and so on, and they're reinforced by allowing them to intrude. By contrast, there is a way the *physical* world is. That way has nothing to do with our needs, desires, tastes, or limitations. A clear-eyed and rigorous interpretation of fundamental physics, unadulterated by consideration of faddish contingencies, will reveal that way. And an adequate philosophy of science will recognize the importance of this—and so foreground realism, representation, "getting the ontology right," both in itself and as a success condition for other varieties of scientific achievement, such as explanation, modeling, etc.

It is in keeping both with pragmatist method and with the spirit of fruitfulness to assess the Perennialist's Apology by examining a putatively fundamental physical theory in search of its unadulterated and univocal account of what the world is like. Next I relate the results of just such an investigation.

7.4.2 Interpreting QM_∞

Starting with the perennial question of what apparently fundamental physics represents, (Ruetsche 2011) develops and defends a pragmatist position in the general philosophy of science. The book argues that engaging with the foundations of QFT motivates a way of thinking about the content of physical theories that challenges perennialist expectations. On this way of thinking, physical content is not rigidly fixed once and for all by an uncompromising exoskeleton of theoretical law. Rather, although constrained by a looser set of constitutive theoretical principles, physical content is to a certain extent the creature of the aims, needs, and limitations of theory-users. What it is *in full detail* to be a QFT varies as function of the circumstances in which the theory is applied, and varies in response to our aims and our background knowledge, our scientific aspirations, our broader explanatory commitments, our technical resources. The best way to make sense of QFT is to pragmatize its content.

A small amount of mathematical physics frames the argument for pragmatizing content. Start by some contrasts between theories of classical physics and quantum theories. In the simplest case of a single particle of mass m traveling in one dimension, classical physics specifies the particle's state by assigning it precise position and momentum values. All its other properties are functions of these values—the particle's kinetic energy, for instance, is its momentum squared, divided by $2m$. Hence, the particle's classical state enables us to predict the values of all other physical properties with probability 1. Interrelations between these properties reflect the laws and symmetries of the theory; a mathematical object known as the *Poisson bracket* expresses these interrelations.

Now, the quantum theory of our particle: its state is a vector in a vector space; position, momentum, and other properties correspond to mathematical objects called operators on that vector space. The state vector fails to fix the values of nearly all of these properties. Instead, via a recipe known as the Born Rule, the quantum state defines a probability distribution over possible values of quantum properties. For a given pair of properties, there is typically the following sort of tradeoff: the more accurately the state vector predicts the value of one property, the less accurately it predicts the value of the other. A mathematical object known as the *commutator bracket* sets the terms of this tradeoff. The commutator bracket also structures the collection of quantum properties in a way that reflects the quantum theory's laws and symmetries.

Their manifest differences notwithstanding, there are striking similarities between quantum and classical theories. At the core of each resides a structuring of physical magnitudes, enforced by the Poisson bracket in the classical theory and by the commutator bracket in the quantum theory. This suggests a recipe that, given a classical theory, produces a quantum theory based upon it and known as the classical theory's *quantization*. This

Hamiltonian quantization recipe asks the aspiring quantum mechanic to take the Poisson bracket between the classical position and momentum magnitudes, and try to find a vector space on which act a pair of operators satisfying a commutator bracket of the same form. The aspiring quantum mechanic thereby constructs a vector space *representation of the canonical commutation relations* (or CCRs) defining the quantum theory she seeks. That quantum theory is *the* quantization of the classical theory she started with. Embarking from a representation of the CCRs, she can readily construct the remainder of her quantum theory. Casting the operators furnishing the representation as quantum mechanical position and momentum magnitudes, she can take products and linear combinations of them to build a host of other quantum magnitudes standing to one another in functional and nomic relationships. With this collection of quantum magnitudes at hand, she can define a family of quantum states as those that assign well-behaved probabilities to possible values of those magnitudes.

So far, so good. But what if it were possible, starting from the *same* classical theory T, to follow this Hamiltonian quantization recipe to construct *different* quantum theories? Then we wouldn't really know which of those were "*the* quantization of T." That supposedly precise physical theory would be at best ill-defined and at worst inconsistent!

The Stone–von Neumann theorem is a mathematical result assuring us that unsettling non-uniqueness does not afflict products of the Hamiltonian quantization recipe—provided that recipe embarks from a classical theory for a system with finitely many degrees of freedom (finitely many particles moving in finitely many dimensions, say). No matter how different a pair of representations of the CCRs quantizing such a theory might seem superficially, those representations will always prove to be notational variants on one another. They'll agree about what's physically possible, as well as about what structures of properties physical possibilities instantiate. When a classical theory is suitably finite, the Stone–von Neumann theorem attests, its quantization is essentially unique.

Classical field theories aren't suitably finite. They concern systems like fields, systems whose states are specified (in the most straightforward case) by assigning a number (the field's amplitude) to each point of space. Since there are infinitely many points of space, a field has infinitely many degrees of freedom. We can still follow the Hamiltonian quantization recipe to quantize a classical field theory. We obtain a quantum field theory. But not a unique one. Outside the scope of the Stone–von Neumann theorem, there are infinitely many apparently physically distinct ways to quantize a given classical field theory. Distinct quantizations can differ on such physically basic questions as what particles are, whether they exist at all, and if they do, whether it's possible to have only finitely many of them.

With good old-fashioned ordinary quantum mechanics, we at least know what vector space structure a theory has. (We just didn't know how to make sense of it!) With QFTs and other quantum theories addressing infinite systems (all of which I gather under the heading QM_∞), there are uncountably many rival vector space structures, affording infinitely many apparently physically distinct representations of the CCRs defining each theory. Each representation gives rise to a quantization that seems qualified to serve as that QFT. Such rampant non-uniqueness raises a host of foundational question: what is a quantum theory? What criteria of identity apply to such theories?[5] What does it really take to be a quantum state or a quantum property?... And: how should we frame and adjudicate answers to questions such as the foregoing?

Two broad strategies of response to the non-uniqueness suggest themselves immediately. The *privileging strategy* identifies the QFT with a unique physically significant vector space representation of the CCRs. Rival representations are dismissed by the privileging strategy as mathematical artifacts without physical significance. The *abstraction strategy* identifies the QFT with a structure common to *all* representations of the CCRs. Features parochial to particular representations are dismissed by the abstraction strategy as superfluous mathematical structure without physical significance. Ruetsche 2011 examines uses to which theories of QM_∞ are put, in the hopes that a winning interpretive strategy, a strategy that makes the most sense of the most uses, will emerge. Such an examination is maximally hospitable to the prospects for fundamentalism/realism. Those positions ask us to believe the world is the way fundamental physics says it is. A winning interpretive strategy would not only tell us what we believe when we believe this, but also *why* we should believe it. Our belief explains the myriad successes of the theory we're interpreting, and thereby enjoys tremendous abductive support. A winning interpretive strategy equips the theory both with unpragmatized content and reason to endorse that content.

[5] Questions about *what theory to interpret* can be distinguished from questions about *how to interpret* a fixed theory—with perennialist and pragmatist answers available, arguably independently, to questions of each sort. This is an important subtlety I suppress in the text, which treats the questions at hand as questions about how to interpret a fixed theory.

The fundamentalist hope for a winning interpretive strategy is dashed, or so Ruetsche 2011 contends. Theories of QM_{∞} are used in many contexts: particle physics, cosmology, black hole thermodynamics, solid state physics, homely statistical physics; and with many aims: to model, explain, predict, and guide the development of future physics. An interpretive strategy that secures one aim in one context may frustrate another aim in another—or even in the same—context. Some examples: The privileging strategy has purchase in standard particle physics, where the symmetries of a particularly simple spacetime (Minkowski spacetime) furnish principles for selecting a privileged representation, one which sustains a fundamental particle notion. However, some aspects of standard particle physics, such as the "soft photons" involved in certain scattering experiments, escape the reach of the privileged representation—but can be modeled by discarded representations. Moreover, significant explanatory agendas invoking particles can't operate within a single privileged representation. Cosmological particle creation is an example. Accounts of this phenomenon deploy rival representations to describe the universe at different epochs in its history. And there are in QM_{∞} other explanatory agendas—embracing symmetry breaking, phase transitions, superconductivity, the dynamics of an expanding universe—that invest a variety of representations with physical significance. The privileging strategy would block these explanatory agendas.

To be sure, some of the explanatory agendas undermined by the privileging strategy are supported by the abstraction strategy. But only some of them. Among the "surplus" properties the abstraction strategy deprives of physical significance are the order properties that distinguish between the distinct phases in a phase transition, as well as the properties that enable us to make sense of the dynamics of mean field models. So there are worthwhile physical projects promoted by each strategy, worthwhile projects frustrated by each strategy, and worthwhile physical projects frustrated by both strategies.

A winning strategy for interpreting QM_{∞} has failed to emerge. Does it follow that we don't understand QM_{∞}? On the contrary, or so I would contend. Noticing the failure—noticing that equipping a theory of QM_{∞} with constitutive CCRs leaves open a host of interpretive questions, questions which can be and in practice are answered in different ways in different contexts of aim and application—*is* understanding QM_{∞}.

Focusing on the fully rigorous theories of mathematical physics afforded by QM_{∞}, regarding them as possible representations of the way the world is, articulating those representations in order to exploit standard realist arguments to justify belief their accounts of the physical world—in short, operating in a manner that seems maximally hospitable to finding perennial answers to perennial questions—I conclude that theory specification has a pragmatist dimension. Strategies for equipping physical theories with content, if sensibly pursued, eventuate in contents indexed not only (or not just) to the way the world is, but also to our aims in using theories and the circumstances we use them in.

7.5 Future Physics?

7.5.1 More Apologetics

Although based on consideration of clean and precise theories of physics, the foregoing move to pragmatize theoretical content is subject to a variation on Perennialist's Apology. Clean and precise though they may be, the perennialist might venture, the theories of QM_∞ I consider are not final physics and they are not fundamental physics. They are ephemeral and imperfect stopgaps. Final physics, when we have it, won't be prone to or compromised by the constrained ambiguity I suggest afflicts present theories of QM_∞.

There is very obviously potential here for a dialectical impasse. Confronted with the Perennialist's Apology, pragmatists might object that the posited fundamental theory underwriting the Apology is nowhere near to at hand. Even bracketing the perennial sticking points of consciousness and the direction of time, the physical theories currently available can't account for most of the gravitational sources in the universe; confining attention to the sorts of matter they can account for, they can't explain why that matter predominates over anti-matter and they can't unite gravity with the forces that fall under the quantum mechanics of the Standard Model. (And that's just the start of their shortcomings.) Perennial philosophers may remain unmoved by these observations, and invoke future theories they can imagine but not produce, theories whose supposed features will neutralize pragmatist conclusions.

The only way I can see to navigate the impasse is to find grounds for expectations about future theories. While such grounds will inevitably be defeasible (or Nobel-prize worthy), the projects of developing and defeasing them are liable to render the perennial and pragmatist theses more contentful, as well as to clarify what's at stake in choosing between them. So: what, in spite of our current ignorance, can we reasonably expect for the future of putatively fundamental physics?

There are forcefully articulated positions in the literature defending the anti-fundamentalist prediction that putatively fundamental physics will never reach a state of univocal completion. This is one way to read Wilson on the ubiquity of physics avoidance (2017) and Cartwright on the untidyness of nature (1999). Wilson develops natural histories of concepts in applied mathematics; Cartwright considers the diverse methodologies of many sciences to evoke the metaphysics of a dappled world. What I'm going to try to do here is mount a *different* defense of the anti-fundamentalist prediction, a defense that concedes as much as I can bear (and possibly more than is reasonable) to fundamentalists and other perennialists.

So although I'm personally attracted by dappled world metaphysics—I like the idea that the world is fundamentally untidy—I'm going to grant both that there's a univocal fundamental way the world is and that there's a possible scientific theory—I'll call it *T*! – that completely and adequately captures that way. Believing *T*!, you'd be believing the truth. In the face of lessons pragmatists would draw from ephemeral science, the Perennialist's Apology asserts that a wholly perennial natural philosophy would result from properly analyzing *T*!. So understood, the Apology incorporates an undischarged promissory note to the effect that *T*! admits an unadulterated, univocal analysis congenial to the aims of perennial philosophy. I'm going to grant that too: I'm going to grant that *T*!' s myriad applications can all be sustained by a single, unambiguous account of what the world is like, according to *T*!.

What I'm going to try to add to the debate are considerations from the past and present of science that support a prediction and a claim. The prediction is that *T*! does not lie in the future of science. The claim is that successful present science is no guide to *T*!. I mean the prediction and the claim to suggest (although they fall a long way short of establishing) that conjuring *T*! and its pristine interpretation won't help us understand the ongoing human effort that is science. Conceding the conceptual possibility of perennialist natural philosophy, I'm nevertheless predicting that it will never pass muster as a *philosophy of science*.

7.5.2 The Jungle Red in Tooth and Claw

As a defense against the pragmatist charge that the content of QFT has a pragmatic dimension, the Perennialist's Apology conjures a final fundamental theory *T*! whose content needn't be pragmatized to make sense of its empirical success. *T*!' s unpragmatized content is simply and straightforwardly a representation of how the world is.

I doubt that the future of human science will comprehend *T*!. Here I defend the anti-fundamentalist prediction that science will never stabilize to a univocal final theory whose content is best understood as unpragmatised. I can't offer conclusive support for the prediction. (Nor can fundamentalists offer conclusive support for the prediction that *T*! waits in the offing. Short of producing *T*!, I'm not sure what would count.[6]) But I can offer support for the anti-fundamentalist prediction in the form of a commentary on a notorious metaphor from van Fraassen's *Scientific Image*. He is countering the realist's contention that the approximate truth of established scientific theories is the best explanation of their remarkable success:

> I claim that the success of current scientific theories is no miracle. It is not even surprising to the scientific (Darwinist) mind. For any scientific theory is born into a life of fierce competition, a jungle red in tooth and claw. Only the successful theories survive—the ones which in fact latched on to actual regularities in nature. (van Fraassen 1980, 40)

My commentary is that the sort of semantic indecision that enables QM_∞ to admit a variety of interpretations is an underappreciated scientific virtue. Scientific theories whose contents are pragmatized exhibit, in the range of interpretations they admit, a resource of constrained adaptability. Admitting different interpretations in different contexts, QM_∞ is equipped to compete in the scientific jungle red in tooth and claw—to meet the demands, many and varied, a living scientific theory faces.[7]

The commentary inspires a prediction, one that can be falsified by the future of science. The prediction is that because pragmatized content is a scientific resource, successor theories will share with QM_∞ the feature that no single interpretation emerges as the best. To do the work the Perennialist's Apology demands of it, *T*! must be free of constrained ambiguity. So put another way, the prediction is: *T*! won't be among the sequence of successor theories. Semantic indecision isn't a passing frailty of present science but a critical strength of science as humans practice it.

[6] Well, I guess they could argue for something like Peirce's analysis of truth—but then they'd still be pragmatizing *something*!

[7] A thought I lack space develop here: there's a connection between the virtue of constrained ambiguity and the fruitfulness norm. Research designs that permit constrained ambiguity may allow evidence to register in a wider range of keys than would research designs that rigidly lock theories into unitary and perennial interpretations.

7.5.3 The Physics of Ignorance

The last section made a positive case for the anti-fundamentalist prediction that univocal final fundamental theory $T!$ does not lie in the future of science. This section considers, and seeks to defuse, a case that we *already* know some things about $T!$. The case rests on the claim that we have on hand resources for identifying which features our present successful theories share with $T!$. These features are prima facie candidates for representations of how the physical world is, and so suitable elements of perennial natural philosophy. The claim that we can, right now, identify features that will persist throughout the future of physics invokes what are known as renormalization group considerations.

7.5.4 A Puzzle about Perturbative QFT

Can we ever tell when our favored theories latch on to the way the physical world is? A position I'll call *renormalization group realism* is a (QFT-specific) strategy for articulating a realism evolved to resist the Pessimistic Meta-induction, a strategy rooted in a response to a foundational question about interacting QFTs. Identifying which aspects of our current theories *represent the world as it is*, this strategy promises to inoculate some theoretical contents against pragmatist encroachment.[8]

Whereas the mathematically tractable QFTs most amenable to the abstraction strategy discussed in the last section concern non-interacting quantum fields, the wage-earning QFTs making up the Standard Model of particle physics deal with interacting fields. The subsequent non-linearity blocks the standard construction of a fully explicit vector space representation for these QFTs. Physicists settle instead for a host of approximation techniques that succeed in extracting confirmable predictions from elements of the theoretical framework. The *perturbative* approach to interacting QFT frames many of these techniques.

In the perturbative approach, an interacting QFT that we don't understand mathematically is modeled as a slight modification of free theory that we do understand mathematically. Here, "understand mathematically"

[8] The position I'm calling renormalization group realism is a stock version of much subtler and thorough views presented in Fraser (2020ab) and Williams (2019, see also 2021). Wallace (2006) is an outstanding discussion of the foundations issues animating the view.

means having an explicit space of solutions to the equations of motion defined by the theory's Lagrangian, and "slight modification" means the full Lagrangian of the interacting theory is a sum of the Lagrangian L_0 of the free theory and a Lagrangian L_I, describing an interaction, multiplied by a coupling coefficient λ assumed to be small:

$$L_{tot} = L_0 + \lambda L_I$$

The ϕ^4 theory is the simplest case of an interacting QFT. It describes a self-interacting mass m scalar field theory, and its total Lagrangian is a function $L(\phi, m, \lambda)$ of that field ϕ, its mass m, and the coupling λ.

We don't know how to solve the equations of motion determined by the total Lagrangian. We do know that those determined by the free Lagrangian L_0 have solutions $\phi_0(x)$. Given the assumption that the coupling λ is small, it is reasonable to hope that solutions $\phi(x)$ for the total theory take the form of a *perturbative expansion* around free solutions:

$$\phi(x) = \phi_0(x) + \lambda\phi_1(x) + \lambda^2\phi_2(x) + \ldots$$

If such solutions are brought under perturbative control, empirically significant quantities—for instance, n-point functions, which foster the calculation of S-matrix elements, which give probabilities for scattering experiments—can be derived.

(One!) problem physicists pursuing this program encounter is that *individual terms* in the perturbative expansion *diverge* (that is, become infinite). The ϕ^4 self-interaction of a mass m scalar field in 4 spacetime dimensions is an example. There, integrals like

$$\int_0^\infty \frac{d^4k}{k^4 + m^2}$$

(k momentum) crop up in individual terms of the perturbative expansions for quantities of empirical interest. Note that the integral ranges over *all* apparently physically reasonable values of momentum, from 0 to ∞.

Physicists have evolved *perturbative renormalization* techniques $\int_0^\infty \frac{d^4k}{k^4+m^2}$ to tame these disruptive infinities. The original problem that integrals like $\int_0^\infty \frac{k^{d-1}}{k^2+m^2} dk$ diverge, rendering the theory empirically meaningless, has a fix known as "regularization." A regularization procedure alters the troublesome divergent integral to make it finite. For example, a regularization procedure might impose an *ultraviolet cutoff*: instead of integrating over momenta from 0 to ∞, truncate the integral at some finite maximum value Λ_{UV} of momentum:[9] $\int_0^{\Lambda_{UV}} \frac{k^{d-1}}{k^2+m^2} dk$. The cutoff integral is finite, which cures the divergence.

But it introduces a new problem: now the theory is cutoff dependent! It doesn't reflect just how the world is, but also an apparently arbitrary choice of a limit of integration. The fix for this is called "renormalization." The basic idea is to reparameterize, that is, rewrite the original Lagrangian in terms of new coefficients cleverly chosen so that as $\Lambda_{UV} \to \infty$, terms in the perturbative expansion at each order converge, and do so in such a way that cut-off dependence disappears. Brought home to the ϕ^4 theory, renormalization replaces our original total Lagrangian $L(\phi, m, \lambda)$ with a *renormalized* Lagrangian $L(\phi_r, m_r, \lambda_r)$ that's a function of masses, couplings, and fields (m_r, λ_r, ϕ_r) different from those (m, λ, ϕ) appearing in the original Lagrangian. As long as the renormalized Lagrangian depends on only finitely many terms—that is, as long as it can be obtained by means of *only a finite number* of reparameterizations of quantities in the original Lagrangian—the physics it defines is humanly tractable. In this case, the theory is lauded as *perturbatively renormalizable*. Think of this as tamable by daredevil redefinitions and fancy calculational footwork. Typically, renormalized masses and couplings aren't supplied by calculations mediated by specific (and possibly inapt) renormalization schemes but *determined experimentally*.

We've taken enough on board to frame a puzzle about perturbative QFT. James Fraser puts it well:

> The success of the perturbative approach is mysterious, I suggest, precisely because it dodges the question of what an interacting QFT is.... [There is] an absence of any non-perturbative characterization of the system of interest. While I have argued that this does not render perturbative QFT incoherent it undercuts the possibility of telling a physical story which could explain its success. (Fraser 2020a, 17–18)

The puzzle is: *Why does the technique of perturbative renormalization work so well*? Renormalization Group (RG) analyses suggest a solution.

[9] An upper bound on momentum translates, via the deBroglie relations, to a lower bound on frequency; hence the nomenclature "ultraviolet."

7.5.5 Renormalization Group Considerations

This thumbnail sketch of RG approaches will associate a physical theory with a Lagrangian L, understood as a function of some list of fields and their space-time derivatives. We might be especially interested in what that Lagrangian implies about phenomena accessible at some scale μ (e.g., an energy achievable in a particle accelerator)—what predictions the Lagrangian underwrites about experiments conducted at that scale. Introduce the notation L^{μ} for a Lagrangian, *taken as* governing phenomena at scale μ. L^{μ} is an element of a space I'll call $\mathcal{T}$. $\mathcal{T}$ has the structure *Lagrangians* × *scales*—it's a space of various Lagrangians, understood as governing physics at various scales.

Where ℓ is the energy scale accessible in our best particle accelerators and Λ is a *more fundamental* (i.e., higher energy) scale, an enticing element of $\mathcal{T}$ is a Lagrangian L^{Λ} governing higher energy physics underlying physics at "our" scale ℓ. It is quite natural to be curious about what L^{Λ} implies about physics within our experimental reach. We can introduce a gadget $R_{\ell\Lambda}$ that acts on $\mathcal{T}$ to track the implication:

$$L^{\ell} = R_{\ell\Lambda} L^{\Lambda}$$

L^{ℓ} is the *effective* (at scale ℓ) Lagrangian induced by the underlying (at scale Λ) Lagrangian L^{Λ}. L^{ℓ} encapsulates everything the underlying Lagrangian implies about physics at our scale. The family of transformations $R_{\ell\Lambda}$ acting on $\mathcal{T}$, the family of transformations that encapsulate the implications for effective physics at one scale of Lagrangians at underlying scales, constitutes the "*renormalization group*" (RG).

This framework suggests the following way of thinking about perturbatively renormalized QFTs: a renormalized Lagrangian obtained through "suspicious manipulations on formal power series" (Gawedzki 1986, 1280) codes a low-energy effective (at scale ℓ) theory induced by (unknown) higher energy theories, and connected to them by an RG flow. Now suppose that the RG's action on the space $\mathcal{T}$ containing our renormalized Lagrangian has the following features:

- *All* high energy theories flow to the *same* subspace of $\mathcal{T}$—the subspace $\mathcal{T}^{\ell}_{eff}$ of effective (at ℓ) theories.
- This *surface of attraction* $\mathcal{T}^{\ell}_{eff}$ is *finite dimensional*: only finitely many parameters are required to specify a Lagrangian in $\mathcal{T}^{\ell}_{eff}$.

[i] would make the details of the underlying theory irrelevant to the shape of the effective theory: even if we're ignorant of the finer points of higher-energy physics, we know in outline what signature they leave on physics at our scales. [ii] would ensure that that signature reduces to a set of experimentally tractable couplings, the parameters necessary to specify the element of $\mathcal{T}^{\ell}_{eff}$ that is the effective theory describing actual physics at scale ℓ.

Effective theories residing in finite dimensional surfaces of attraction have the virtue that the contribution of unknown physics to physics at the scales we care about can be fixed experimentally, simply by measuring coefficients (masses and couplings and the like) defining the effective Lagrangian. Empirically successful perturbatively renormalizable interacting QFTs are widely regarded to be effective theories of exactly this sort:

> It is a remarkable property of local QFTs that for a certain subset of theories,...the low energy amplitudes can be parameterized by just a finite set of parameters—namely, those needed to locate the theory on the finite-dimensional attractive submanifold, and which can in principle be determined by making an equal number of independent experimental measurements. (Duncan 2012, 587)

Polchinski (1984) establishes one instance of the "remarkable property" Duncan celebrates. Polchinski considers a space of theories $\mathcal{T}$ consisting of Lagrangians that are polynomials in a 4 (spacetime) dimensional scalar field and its derivatives, and that take the form of the free Klein-Gordon field with *weakly coupled* interactions. He develops a precise definition of the RG map on that space, and shows that under the action of RG flow induced by this map, $\mathcal{T}$ has a 3 dimensional surface of attraction, where this surface is parameterized by couplings m^* (the effective mass), λ^* (the effective coupling) and ϕ^* (the effective field). This surface includes the perturbatively renormalized ϕ^4 Lagrangian we know and love!

Perturbative QFT's real problem was: why should the approximation technique of perturbative renormalization work at all? What are the approximations approximations to? The answer suggested by RG analyses is: perturbatively renormalized Lagrangians, astonishingly successful at applications to experimentally accessible regimes, are the effective Lagrangians for finitely renormalizable theories induced by physics underlying *physics at (experimentally accessible) scale ℓ*. This underlying physics induces tractable effective theories because physics at scale ℓ requires only finitely many renormalized parameters to specify. Put in terms of the RG apparatus: $\mathcal{T}^{eff}_{\ell}$ is finite dimensional and parameterized by the very variables perturbative renormalization techniques manipulate to tame unwieldy infinities. Despite not knowing the high energy details, we can completely specify the physics at scale ℓ by *measuring* the finitely many couplings that locate the effective theory in $\mathcal{T}_\ell^{eff}$. So we'll get the effective physics right even if our perturbative renormalization schemes are opportunistic and our cutoff procedures ill-understood.

7.5.6 Renormalization Group Realism

One strategy by which the realist might respond to the pessimistic meta-induction is *divide et impera*: distinguish aspects of our best current theories we can reasonably expect to persist in the face of theory change from idle theoretical wheels, and espouse realism about the distinguished aspects. For the strategy to amount to a move (rather than a promise or a bluff), the realist needs a way to pick out the distinguished aspects. The approach just sketched not only suggests a solution to the foundational puzzle about perturbative renormalization techniques. It also suggests a more precise *divide et impera* strategy than others currently on offer, a strategy undergirding a position I'll call *Renormalization Group Realism.*

Renormalization Group Realism acknowledges our most successful interacting QFTs to be merely effective. They succeed at experimentally accessible low energy scales because they approximate the implications higher-energy physics (at present unknown) holds for phenomena at those scales. Nevertheless, RG considerations enable us to identify features of these effective theories that are robust under variations in the underlying, unknown higher energy physics. These are features it is reasonable to expect to persist through future refinements of physics, features apt for a realism resistant to the pessimistic meta-induction, features in that sense perennial. They're features that promise to equip the physics we have now with unpragmatized content.

While I endorse the appeal to RG considerations as a resource in the project of making sense of QFT's empirical success, I also question whether the RG Realist strategy just outlined establishes a viable bulwark against attempts to pragmatize content. Here I'll argue that, cast as a response to the pessimistic meta-induction, RG Realism faces a dilemma. Either the strategy is merely a hopeful metaphor, or it's vulnerable to the anti-realist maneuver of conjuring future physics that lies outside the space on which the RG acts. I'll try to drape the second horn in some historical credibility by casting Newton's LUG as an effective theory.

We can use our old friend, unknown fundamental theory $T!$, to characterize what RG realism offers the perennialist. RG realism offers the perennialist a way to identify elements of present effective physical theory T that T shares with $T!$. *No matter what* $T!$ *is,*"robust" features of T, identifiable by appeal to RG considerations, will be retained its successors; these features are reliable (albeit partial) guides to $T!$ and thus enduring (albeit partial) answers to the perennial question of what the physical world is like. But this is overstating what RG considerations can secure. RG considerations secure the persistence of "robust" features on the hypothesis that the space $\mathcal{T}$ of theories on which the RG acts includes $T!$. Existing RG results—of which Polchinski's is an unusually explicit example—exploit carefully circumscribed spaces of underlying theories. Mathematical physicists recognize that the circumscription calls for circumspection:

> Needless to say that we do not know how to solve [the RG equation] in general. *Some approximations are thus required....* the strategy consists in *solving the RG equation in a restricted functional space* and not as a series expansion in a small parameter. This is why we can hope to obtain non-perturbative results.... Of course, *the quality of the result will depend crucially on the choice of space* in which we search for a solution.... In all cases, *it is impossible to know whether we have missed some physically crucial ingredient by making one choice rather than another one.* (Delamotte 2007; emphasis added)

Even explicit RG results are only as reassuring as the space of theories on which the RG group acts is comprehensive. But that space incorporates assumptions about what kinds of interactions are possible and how to model them. And nature isn't beholden to obey those assumptions. $T!$ could lie outside $\mathcal{T}$ and beyond the reach of the RG. If so, RG considerations can't assure us that any part of present effective theory T reliably reflects either $T!$ or the truth about the physical world.

The RG realist faces a dilemma. Either the space of theories $\mathcal{T}$ on which the RG acts is explicitly specified, or it isn't. If $\mathcal{T}$ is specified, the Pessimistic Meta-Induction retreats a level: skeptically relevant $T!$s $\notin \mathcal{T}$ undermine grounds for belief in T's representational success. If on the other hand, $\mathcal{T}$ isn't specified, neither are reassuring RG results about which features of T are invariant under the details of the underlying physics.

A slightly different (and imperfectly analogous!) context supplies a concrete example of the epistemic dangers of overconfidence in the comprehensiveness of a theoretical space (see Wells 2012 and Ruetsche 2018 for the example developed). Cast Newton's Law of Universal Gravitation (LUG) ($V(r) = \frac{GMm}{r}$) as a free theory residing in a space $\mathcal{T}$ of perturbative corrections. A generic member of this space takes the form:

$$V(r) = \frac{GMm}{r}[1 + \sum_{n=1}^{\infty} \lambda_n (\frac{r_{0_n}}{r})^n]$$

LUG isn't exactly true: famously, it fails to accommodate the perihelion advance of Mercury and other planets. However, other elements of $\mathcal{T}$, e.g., the cubic correction $V_2(r) = \frac{GMm}{r}(1 + \frac{R_2^2}{r^2})$, can save the observed phenomena of perihelion advance, provided finitely many free parameters are fixed by experiment.

Such reflection on the space $\mathcal{T}$ containing LUG and perturbative corrections thereto might inspire a kind of optimism: no matter what the details of underlying fundamental physics, its implications for physics at orbital radii we care about can be captured by a finite number of small empirically tractable corrections to LUG. In this sense, *LUG is approximately true,* and we can with confidence subscribe to those features of LUG's picture of the world that it shares with other theories in $\mathcal{T}$. These include central gravitational forces acting instantaneously at a distance across Euclidean three-space.

But hindsight is 20:20, and in this case, pessimistic. The advent of General Relativity (GR) supports a starkly different commentary on LUG as an effective theory. The equations of motion determined by the cubic correction to LUG are low velocity, low eccentricity, large radius limits of the equations of motion arising from the Schwarzschild solution of the field equations of GR. GR does away with central forces acting instantaneously across Euclidean space. Reassuring considerations of features it shares with an apparently encompassing space of theories notwithstanding, *LUG gets things dead wrong.* Taking features of LUG "robust" under incorporation of higher-order corrections to represent how the world is, we would have been led astray.

The analogy, between RG realism and realism about what LUG shares with its space of perturbative corrections, is imperfect. Still this example should shake our faith in guides to the unpragmatized content of final fundamental physics that rely on examination of a supposedly complete space of possible theories. And it should do something else. It should help us to recognize a tension between research approaches framed by a posited "space of theories" (the $\mathcal{T}$s on which the RG acts, the space of perturbative corrections to LUG, and string theory's "landscape" are all prima facie examples) and the methodological norm of "fruitfulness." Resolving to confront phenomena within the ambit of the space of perturbative corrections to LUG, physics would never have recognized anomalous perihelion advance as evidence for a theory, General Relativity, operating outside of that ambit.

7.6 A Pragmatist Reappropriation

RG realism's robustness criterion picks out *some* content for endorsement. The last section questioned whether the content selected merited perennial endorsement as a reliable representation of how the physical world is. This section asks: if the robust content doesn't deserve perennial endorsement, what kind of endorsement does it deserve? My closing suggestion will be that the endorsement it deserves has a strongly pragmatist flavor.

Features of theories anointed as robust by RG considerations aren't thereby declared to be reliable representatives of fundamental physical reality. They aren't certified as real *tout court*. They are rather certified to be features that will persist as elements of theories *effective at scale* ℓ, no matter what physics reigns at higher energy scales. Both the *scale-dependence* of the content endorsed by robustness considerations and their endorsement of that content as *effective* leave RG realism vulnerable to pragmatist reappropriation. A pertinent value for ℓ when applying RG considerations is one that reflects the energies we can access experimentally. The content RG realism endorses is thus sensitive to our experimental reach, and liable to vary as our technologies evolve. Indexing theoretical content to our present empirical and technological capacities is one way to pragmatize that content.

But RG realism harbors an even more dramatically pragmatic aspect. Contrasted with the sort of perennialism Maudlin's naturalistic metaphysics exemplifies, RG realism advocates a seismic shift in terms of theoretical endearment. Whereas Maudlin celebrates physics qualified to inform metaphysics as "fundamental," RG realists hail our best QFTs as "effective." Rather than a compendium of nature's building blocks or a catalog of some ultimate supervenience base, an effective (at scale ℓ) theory is a successful coping mechanism, a reliable means of organizing phenomena that manifest (at scale ℓ). That is, effective theories abound in *pragmatic* virtue, with those elements of effective theories endorsed by the robustness criterion exhibiting the additional pragmatic virtue of probable continued utility. Whether effective theories are *true*, or harbor *genuinely referential* elements, where those accomplishments demand something more than the pragmatic virtues just enumerated—the answers to these questions don't influence the patterns of endorsement issued by RG realism. It could well be that RG realism is pragmatism in denial![10]

References

Anderson, E. (2004). Uses of value judgments in science: A general argument, with lessons from a case study of feminist research on divorce. *Hypatia, 19*(1), 1–24.

Carnap, R. (1950). Empiricism, semantics, and ontology. *Revue internationale de philosophie, 4*(11), 20–40.

Cartwright, N. (1999). *The dappled world: A study of the boundaries of science.* Cambridge University Press.

Delamotte, B. (2007). An introduction to the nonperturbative renormalization. *grouparXiv*. arXiv preprint cond-mat/0702365.

Duncan, A. (2012). *The conceptual framework of quantum field theory*. Oxford University Press.

van Fraassen, B. (1980). *The scientific image*. Oxford University Press.

Fraser, J. D. (2020a). The real problem with perturbative quantum field theory. *The British Journal for the Philosophy of Science, 7*(2), doi.org/10.1093/bjps/axx042.

Fraser, J. D. (2020b), Toward a realist view of quantum field theory. In French and Saatsi, eds., *Realism and the quantum*, Oxford University Press, 276–292.

Hrdy, S. (1986). Empathy, polyandry, and the myth of the coy female. In Bleier, ed., *Feminist approaches to science*. Pergamon, 119–46.

Gawedzki, K. (1986). Renormalization group, from magic to mathematics. Talk delivered at the International Congress on Math (Berkeley).

Maudlin, T. (2007). *The metaphysics within physics*. Oxford University Press.

[10] Elsewhere (Ruetsche 2020), I question whether the attitude emerging from the RG Realist strategy should, on its own, be understood as realism. I argue that the commitments most clearly validated by the strategy are either commitments the anti-realist can live with outright, or commitments that the anti-realist can live with until and unless they are supplemented by something RG realists forswear: a Good Old Fashioned "Standard Interpretation" of QFT, an account of what sort of microworld is truly described/aptly represented by the theory.

Mitchell, S. (2003). *Biological complexity and integrative pluralism*. Cambridge University Press.

Polchinski, J. (1984). Renormalization and effective Lagrangians. *Nuclear Physics, B231*(2), 269–95.

Price, H. (2009). Metaphysics after Carnap: The ghost who walks. In Chalmers, Manley, & Wasserman, eds., *Metametaphysics: New essays on the foundations of ontology*, Oxford University Press, 320–46.

Ruetsche, L. (2011). *Interpreting quantum theories*. Oxford University Press.

Ruetsche, L. (2015). The shaky game+ 25, or: on locavoracity, *Synthese, 192*(11), 3425–42.

Ruetsche, L. (2018). Renormalization group realism: The ascent of pessimism, *Philosophy of Science, 85*(5), 1176–89.

Ruetsche, L. (2020). Perturbing realism. In Steven French & Juha Saatsi, eds., *Realism and the quantum*, Oxford University Press.

Schmitt, C. B. (1966). Perennial philosophy: From Agostino Steuco to Leibniz. *Journal of the History of Ideas, 27*(4), 505–32.

Teller, P. (2012). Modeling, truth, and philosophy. *Metaphilosophy, 43*(3), 257–74.

Wallace, D. (2006). In defence of naiveté: The conceptual status of Lagrangian quantum field theory. *Synthese, 151*(1), 33–80.

Waters, C. K. (2006). A pluralist interpretation of gene-centered biology. In Kellert, Longino, & C. K. Waters, eds., *Scientific pluralism: Volume XIX of the Minnesota studies in the philosophy of science*. University of Minnesota Press, 190–214.

Wells, James (2012). *Effective Theories in Physics: From Planetary Orbits to Elementary Particle Masses*. Springer.

Williams, P. (2019). Scientific realism made effective. *British Journal for the Philosophy of Science, 70*(1), 209–37.

Williams, P. (2021). Renormalization group methods. In *The Routledge companion to philosophy of physics*. Routledge, 296–310.

Wilson, M. (2017). *Physics avoidance: Essays in conceptual strategy*. Oxford University Press.

Woodward, J. (2023). Sketch of some themes for a pragmatic philosophy of science (this volume).

Index

Note: Figures are indicated by an italic '*f*' following the page numbers.

For the benefit of digital users, indexed terms that span two pages (e.g., 52–3) may, on occasion, appear on only one of those pages.